Springer Tracts in Modern Physics
Volume 170

Managing Editor: G. Höhler, Karlsruh

Editors: J. Kühn, Karlsruhe
Th. Müller, Karlsruhe
A. Ruckenstein, New Jersey
F. Steiner, Ulm
J. Trümper, Garching
P. Wölfle, Karlsruhe

Honorary Editor: E. A. Niekisch, Jülich

Now also Available Online

Starting with Volume 163, Springer Tracts in Modern Physics is part of the Springer LINK service. For all customers with standing orders for Springer Tracts in Modern Physics we offer the full text in electronic form via LINK free of charge. Please contact your librarian who can receive a password for free access to the full articles by registration at:

http://link.springer.de/series/stmp/reg_form.htm

If you do not have a standing order you can nevertheless browse through the table of contents of the volumes and the abstracts of each article at:

http://link.springer.de/series/stmp/

There you will also find more information about the series.

Springer-Verlag Berlin Heidelberg GmbH

Physics and Astronomy

ONLINE LIBRARY

http://www.springer.de/phys/

Springer Tracts in Modern Physics

Springer Tracts in Modern Physics provides comprehensive and critical reviews of topics of current interest in physics. The following fields are emphasized: elementary particle physics, solid-state physics, complex systems, and fundamental astrophysics.

Suitable reviews of other fields can also be accepted. The editors encourage prospective authors to correspond with them in advance of submitting an article. For reviews of topics belonging to the above mentioned fields, they should address the responsible editor, otherwise the managing editor. See also http://www.springer.de/phys/books/stmp.html

Managing Editor

Gerhard Höhler

Institut für Theoretische Teilchenphysik
Universität Karlsruhe
Postfach 69 80
76128 Karlsruhe, Germany
Phone: +49 (7 21) 6 08 33 75
Fax: +49 (7 21) 37 07 26
Email: gerhard.hoehler@physik.uni-karlsruhe.de
http://www-ttp.physik.uni-karlsruhe.de/

Elementary Particle Physics, Editors

Johann H. Kühn

Institut für Theoretische Teilchenphysik
Universität Karlsruhe
Postfach 69 80
76128 Karlsruhe, Germany
Phone: +49 (7 21) 6 08 33 72
Fax: +49 (7 21) 37 07 26
Email: johann.kuehn@physik.uni-karlsruhe.de
http://www-ttp.physik.uni-karlsruhe.de/~jk

Thomas Müller

Institut für Experimentelle Kernphysik
Fakultät für Physik
Universität Karlsruhe
Postfach 69 80
76128 Karlsruhe, Germany
Phone: +49 (7 21) 6 08 35 24
Fax: +49 (7 21) 6 07 26 21
Email: thomas.muller@physik.uni-karlsruhe.de
http://www-ekp.physik.uni-karlsruhe.de

Fundamental Astrophysics, Editor

Joachim Trümper

Max-Planck-Institut für Extraterrestrische Physik
Postfach 16 03
85740 Garching, Germany
Phone: +49 (89) 32 99 35 59
Fax: +49 (89) 32 99 35 69
Email: jtrumper@mpe-garching.mpg.de
http://www.mpe-garching.mpg.de/index.html

Solid-State Physics, Editors

Andrei Ruckenstein
Editor for The Americas

Department of Physics and Astronomy
Rutgers, The State University of New Jersey
136 Frelinghuysen Road
Piscataway, NJ 08854-8019, USA
Phone: +1 (732) 445 43 29
Fax: +1 (732) 445-43 43
Email: andreir@physics.rutgers.edu
http://www.physics.rutgers.edu/people/pips/
Ruckenstein.html

Peter Wölfle

Institut für Theorie der Kondensierten Materie
Universität Karlsruhe
Postfach 69 80
76128 Karlsruhe, Germany
Phone: +49 (7 21) 6 08 35 90
Fax: +49 (7 21) 69 81 50
Email: woelfle@tkm.physik.uni-karlsruhe.de
http://www-tkm.physik.uni-karlsruhe.de

Complex Systems, Editor

Frank Steiner

Abteilung Theoretische Physik
Universität Ulm
Albert-Einstein-Allee 11
89069 Ulm, Germany
Phone: +49 (7 31) 5 02 29 10
Fax: +49 (7 31) 5 02 29 24
Email: steiner@physik.uni-ulm.de
http://www.physik.uni-ulm.de/theo/theophys.html

Bärbel Fromme

d–d Excitations
in Transition-Metal Oxides

A Spin-Polarized Electron
Energy-Loss Spectroscopy (SPEELS) Study

With 53 Figures

 Springer

Dr. Bärbel Fromme

University of Düsseldorf
Institute for Applied Physics
Universitätsstrasse 1
40225 Düsseldorf, Germany
E-mail: fromme@uni-duesseldorf.de

Library of Congress Cataloging-in-Publication Data.

Die Deutsche Bibliothek - CIP-Einheitsaufnahme
Fromme, Bärbel:
d–d excitations in transition metal oxides: a spin polarized electron
energy loss spectroscopy (SPEELS) study / Bärbel Fromme. – Berlin;
Heidelberg; New York; Barcelona; Hong Kong; London; Milan; Paris;
Singapore; Tokyo: Springer, 2001.
(Springer tracts in modern physics; Vol. 170)
(Physics and astronomy online library)

Physics and Astronomy Classification Scheme (PACS):
71.70.Ch, 73.20.-r, 71.20.-b, 82.80.Pv

ISSN print edition: 0081-3869
ISSN electronic edition: 1615-0430
ISBN 978-3-662-14685-9   ISBN 978-3-540-45342-0 (eBook)
DOI 10.1007/978-3-540-45342-0

© Springer-Verlag Berlin Heidelberg 2001

Originally published by Springer-Verlag Berlin Heidelberg New York in 2001.

Softcover reprint of the hardcover 1st edition 2001

Typesetting: Data conversion by EDV-Beratung F. Herweg, Hirschberg using a Springer LaTeX macro package
Cover design: *design & production* GmbH, Heidelberg

Printed on acid-free paper   SPIN: 107757879   56/3141/tr   5 4 3 2 1 0

Preface

Since the discovery of the insulating nature of the transition-metal oxides with incompletely filled 3d shells in 1937, the interest in this fascinating class of compounds, in particular in their electronic structure, has never vanished. Presently, research in this field is gaining increasing importance because a detailed knowledge of the oxides' bulk and surface electronic structure is essential for understanding and optimizing the mechanisms relevant to the growing number of technological applications of these materials, which are used in lasers, sensors, and catalysis. In this monograph, the present knowledge about the electronic structure of the monoxides NiO, CoO, and MnO is briefly reviewed, particularly with respect to the 3d electrons, which remain localized at the transition-metal ions because a strong Coulomb correlation prevents them from forming a partially filled 3d conduction band, leading to the insulating behavior.

We have investigated the electronic structure of the monoxides by studying the dipole-forbidden transitions between the crystal-field-split 3d states ("d–d transitions") of the transition-metal ions by means of *spin-polarized electron energy-loss spectroscopy* (SPEELS), with polarized primary electrons and polarization analysis of the inelastically scattered electrons. The d–d transitions are hardly accessible to optical spectroscopies, but can easily be excited by electrons owing to the possibility of electron exchange. SPEELS allows not only an unambiguous proof of electron exchange processes and investigations of their behavior, but also the identification and investigation of other inelastic scattering mechanisms involved in the excitation process – such as dipole and resonance scattering – if the primary energy and scattering geometry are varied, owing to the different dependences of the different scattering mechanisms on these parameters. The knowledge and use of resonant primary energies in union with spin-resolved measurements is found to be essential for the determination of d–d excitation energies. The variation of the scattering geometry provides additional important information, not only about the contributions of the different scattering mechanisms to the scattering or excitation process, but also concerning the assignment of energy-loss peaks to particular d–d excitations and the identification of the d–d transitions of surface transition-metal ions. With this technique, it was possible to measure and assign nearly all sextet–quartet d–d transitions of

MnO; for CoO, some d–d excitations with higher excitation energies than those previously measured have been measured for the first time. Some of the theoretically predicted surface d–d excitations of NiO have been measured, in addition to bulk d–d excitations. After an introduction to the various relevant inelastic scattering mechanisms, the experimental method and its possibilities are described. The results of our investigations concerning transition-metal oxides are presented thoroughly and compared with other experimental and theoretical results.

This work is an updated and slightly extended version of my post-doctoral thesis (Habilitationsschrift), finished at the end of 1998. The experimental results presented and summarized here have been obtained at the Institute of Applied Physics at the Heinrich-Heine University of Düsseldorf. I am deeply indebted to all my colleagues at this institute – it is hard to find somebody here who did not support my work. Nevertheless, some people have to be mentioned in particular: first of all I want to thank Prof. Dr. Erhard Kisker for his continuous interest in the experiments and his support in the solution of many problems, and also Dr. Hildegard Hammer, who was always ready to help, whatever the problem was – often oblivious of time.

Research at universities is not possible without two groups of people: undergraduate and graduate students, and the technical staff. It gives me great pleasure to thank them here. The undergraduate and graduate students who contributed to this work were Bernd Runge, Markus Schmitt, Alexander Hylla, Christian Koch, Rainer Deußen, Thomas Anschütz, Cersten Bethke, Udo Brunokowski, and Michael Möller. Mechanical and electronic problems were always quickly solved by Stefan Manderla, Wilfried Gjungjek-Schützek, and Ulrich Rosowski. Thank you very much!

Collaboration and discussions with scientists from other institutes and universities were very helpful. In particular I would like to thank Prof. Dr. Horst Merz and Dr. Andreas Gorschlüter for their very extensive collaboration in the early stages of the "oxide project". They also supported us later on – for example by donating some crystals. The stimulating and critical discussions with Prof. Dr. Volker Staemmler, Prof. Dr. Hans-Joachim Freund, Dr. Thomas Schönherr, Prof. Dr. John Inglesfield, and Frank Müller opened up many new aspects concerning the physics of oxides. Thank you! I am additionally indebted to Frank Müller and Dr. Andreas Gorschlüter for providing the data presented in Figs. 3.4 and 5.7, respectively.

I would like to thank Dieter for the endless patience he had with me, especially in difficult phases of this project, and which he showed just recently by the time-consuming reading of the manuscript. I thank Cordelia Koppitz for her careful reading of some sections of the manuscript – several grammatical mistakes were eliminated by her.

I am very grateful to the Ministry of Science and Research of North-Rhine Westphalia for granting a Lise-Meitner-Stipendium. Without this financial support during five years it would have been impossible for me to finish this project. I am also indebted to the German Research Society for the financial support of our experiments.

Düsseldorf   *Bärbel Fromme*
November 2000

Contents

1. Introduction

Transition-metal oxides form a fascinating class of compounds with a wide range of technological applications. They are used in catalysis, lasers, and magnetic recording tapes, as well as in sensors for very high pressures and for gases, for example. New applications in magnetic-field sensors, based on the principle of giant magnetoresistance [67], and in transparent transistors, applicable as on-screen electronic devices in displays or cameras [165], were added recently. In addition, they are responsible for the occurrence of high-temperature superconductivity.

The possibility of these various applications of transition-metal oxides derives from an enormous variety of different physical and chemical properties, based on the electronic structures of the bulk as well as the surface of these oxides. Bulk effects, for example, are responsible for the occurrence of many kinds of electric conductivity and magnetic order: whereas the simple monoxides MnO, CoO, NiO, and CuO are insulators with gaps of up to several electron volts, the Cu-oxide-based perovskites such as $YBa_2Cu_3O_{7-\delta}$ and $Bi_2Sr_2CaCu_2O_{8+\delta}$ show high-temperature superconductivity. Semiconducting oxides ($Fe_{0.9}O$), as well as compounds with metallic conductivity like ReO_3, exist [23]. The magnetic order often differs strongly for similar oxides: whereas Cr_2O_3 is an antiferromagnet, the chromium oxide CrO_2, which is also stable and is used in magnetic recording, exhibits ferromagnetic behavior. The monoxides of the 3d series also show antiferromagnetic order, but with strongly differing Néel temperatures [23, p. 134], [3, p. 5]: NiO has a high Néel temperature far above room temperature ($T_N = 523\,K$) and forms therefore a suitable antiferromagnetic pinning layer for giant-magnetoresistance and spin valve devices, used in magnetic-field sensors and magnetic heads for disk systems, respectively [61, 142]. The very similar insulating compound MnO, however, has a much lower Néel temperature of 118 K and is paramagnetic at room temperature. The catalytic behavior of several transition-metal oxides and the occurrence of adsorption-induced changes in the surface conductivity of ZnO, SnO_2, and TiO_2, which are used in gas sensors, are attributed to the electronic properties of the surfaces [75, 173, 174].

For a lot of applications and for the explanation of several phenomena occurring in the transition-metal (TM) oxides, the TM 3d electrons in particular, and their behavior are of central significance. The 3d electrons and

3d holes remain localized at the TM ions in several oxides and show not band-like, but quasi-atomic-like character. In contrast to free atoms or ions, where the d states are degenerate, the d states of the transition-metal ions in the oxides are energetically split owing to the crystal field provided by the surrounding oxygen ions. Transitions between 3d states are of great influence on the optical properties and often determine the color of a compound. For example, the d–d transitions of the Cr^{3+} ions in Cr_2O_3 are responsible for the green color of this compound, used in ceramic glazing for centuries. The equivalent transitions of Cr^{3+} embedded in the different crystal field of Al_2O_3 provide the beautiful red color of ruby and are used for the generation of laser light in the ruby laser [116,117,125,188]. In addition, the energy shift of these transitions due to the pressure dependence of the crystal-field splitting can be used for measurements of very high pressures of up to more than 10^9 Pa [42].

A knowledge of the behavior of adsorbates is essential for the understanding of catalytic processes at the oxides' surfaces. Here the localized 3d states are also of considerable importance, because the crystal-field splitting of surface transition-metal ions, which deviates from that of bulk ions owing to missing oxygen ions, can be used to determine the adsorption sites of molecules: if the adsorbed molecule occupies regular surface sites, it replaces the missing oxygen ions at the surface and the surface crystal-field splitting is found to be altered after adsorption; it becomes similar to the crystal-field splitting of bulk transition-metal ions again. No changes are expected, however, if the adsorbed molecule resides at other sites, for example surface defects. By use of this effect, the adsorption of NO at regular surface sites of NiO(100) was demonstrated; OH molecules, on the contrary, were found to be adsorbed at defects on this surface [44,46,48].

For the origin of high-temperature superconductivity in the Cu–O-based perovskites, not the localization of the 3d electrons, but the interaction of a *localized Cu 3d hole* with a movable O 2p hole, leading to the formation of a so-called Zhang–Rice singlet, seems to be essential [70,216].

In the transition-metal oxides with incompletely filled 3d shells such as Cr_2O_3, MnO, FeO, CoO, and NiO, the localization of the TM 3d electrons is responsible for the insulating nature of the oxides. These oxides were among the first solids found where the band picture fails to describe a wide range of physical properties. From simple band-structure calculations, the 3d states are expected to form an incompletely filled 3d conduction band, leading to metallic conductivity of the oxides similar to that of the corresponding transition metals [84, pp. 184ff.], which is not in fact observed. In fact, these oxides belong to the class of Mott–Hubbard or charge-transfer insulators, because a strong Coulomb correlation prevents the electrons from forming a 3d band and localizes them at the transition-metal ions. The electrons cannot move freely and an energy of several electron volts is needed for electron transfer between neighboring transition-metal ions. It was the discovery of this insulating behavior of the transition-metal monoxides [27] and its contradiction to the results in the young field of band-structure calculations which started the intense investigations into their electronic structure, which have lasted up to the present day. In the last few years especially, the growing interest in sensors and catalysis and, in particular, the discovery of high-temperature superconductivity in the Cu–O perovskites have forced the intensification of studies concerning the bulk and surface electronic structure of these oxides.

In the work described in this book, the electronic structure of MnO, CoO, and NiO was investigated by studying the dipole-forbidden transitions between the crystal-field-split 3d states of bulk and surface transition-metal ions with spin-polarized electron energy-loss spectroscopy (SPEELS) using low-energy electrons with energies of 20–130 eV. In addition, some transitions across the insulating gap and excitations from upper core levels were examined. After a summary of the present knowledge of the electronic structure of the transition-metal monoxides (Chap. 2) the experimental method is described in detail in Chap. 3. However, some introductory remarks, describing, the motivation for the application of energy-loss spectroscopy and of its more sophisticated version using polarized electrons (SPEELS) to the transition-metal oxides, will be presented here.

Despite an often poorer energy resolution, electron spectroscopies with low-energy electrons provide several advantages relative to optical absorption spectroscopy. The primary energy of the exciting electrons is very easily swept over a wide energy range, corresponding to the energy of visible light up to soft x-rays. This opens the possibility to examine d–d transitions, which require excitation energies of less than $\approx 6\,\text{eV}$, as well as excitations from valence band and upper core levels. In optical spectroscopies, this is possible only if synchrotron radiation is used. With electron energy-loss spectroscopy, collective excitations such as plasmons are also observed; phonons and vibronic excitations of adsorbed molecules can be measured, if the energy resolution is high enough. Owing to the low penetration depth of the electrons, electron energy-loss spectroscopy is very surface-sensitive and allows investigations to improve the knowledge of the surface electronic structure, which is essential for the understanding of adsorption and catalysis. In particular, for the examination of dipole-forbidden excitations such as d–d transitions, which are barely excited by photons, electron energy-loss spectroscopy is excellently suitable owing to the "breakdown" of dipole selection rules: apart from electric dipole transitions, other electric multipole transitions and, especially, *excitation by electron exchange* become possible if electron impact is used for excitation.

Generally, electron exchange is a well-known but at present poorly investigated phenomenon in excitations of atoms, molecules, and solids. In particular, the angle and energy dependence of the inelastic exchange-scattering process[1] is usually not clear. Calculations of the exchange-scattering cross sections are difficult, because, in contrast to other inelastic scattering mechanisms, a detailed microscopic description of the electron–target interaction is needed. The reason for the nearly total lack of experimental investigations is found in the extreme difficulty of identifying exchange processes unambiguously, because it is necessary to distinguish between scattered primary electrons and emitted true target electrons in this case. But electrons are distinguishable only if their spin directions are different. Therefore, an unambiguous experimental proof of electron-exchange excitations requires both a polarized primary electron beam and the polarization analysis of the inelastically scattered electrons, which is possible only with the particular kind of spin-polarized electron energy-loss spectroscopy that was used in the work described here, where a polarized primary electron beam is scattered at the target and the energy distribution and polarization of the scattered electrons are measured simultaneously.

Besides the determination of excitation energies, which is also possible by conventional electron energy-loss spectroscopy with unpolarized electrons, and the direct proof of exchange processes mentioned above, SPEELS provides further advantages: if other parameters such as the primary energy or scattering geometry are varied additionally, it is possible to distinguish between different scattering mechanisms such as exchange, dipole, and resonant scattering. These different scattering mechanisms correspond to different interactions between the incident electrons and the target, leading to excitation of the target and the inelastic scattering process. The contributions of the different electron–target interactions can be determined, which allows conclusions about the kind of excitation to be drawn. Their scattering-geometry and energy dependence can be examined.

The experimental setup is briefly described in Chap. 4; details of the crystals used and the preparation of their surfaces are also described there. The results of our investigations of the transition-metal monoxides are presented and discussed in Chap. 5.

---

[1] The terms "exchange scattering" and "exchange excitation" are used in parallel in this work, as in the literature, because they are almost equivalent descriptions of the same physical process: electrons are inelastically scattered at a target, which is excited in the process. If the excitation is accompanied by electron exchange, the scattering process is called "exchange scattering".

2. Electronic Structure of MnO, CoO, and NiO

2.1 Introduction

Investigations of the electronic structure of transition-metal oxides with incompletely filled 3d shells have a long history [15]. With the discovery of their insulating behavior, presented at a conference in Bristol in 1937 [27], they were among the first highly correlated systems found, where band theory fails to describe a wide range of physical properties. For the monoxides it is mainly the large size of the insulating gap and often the occurrence of a gap at all which cannot be explained adequately in one-particle band-structure formalisms such as the local-density approximation (LDA). Determined attempts have been made during the last sixty years to understand the origin of the insulating gaps, and it were the ingenious ideas of Peierls, Wilson, and Mott [137] at the Bristol conference and the later work of Mott [138] and Hubbard [81] which brought the problems closer to a solution. According to these ideas, which led to the development of the concepts of Mott–Hubbard and charge-transfer insulators later on, a strong Coulomb correlation between the d electrons is responsible for the insulating nature of the monoxides. Briefly, the d electrons remain localized at the metal ions, because their Coulomb correlation prevents them from forming an incompletely filled 3d band. They do not behave like a gas of easily movable electrons, because an energy of several electron volts is needed for the transport of electrons through the crystal lattice, and the conductivity is therefore very low.[1]

In addition to the origin of the insulating gaps, several experimental results obtained from optical and electron energy-loss spectroscopy cannot be understood in terms of band-structure calculations, because they clearly reflect the localization and correlation of the 3d electrons. For example, only a very small dispersion of the 3d states is found in angle-resolved photoemission spectra [82, 83, 105, 113, 177, 178, 179], and excitations from core levels into 3d states – in particular the 3p–3d excitations – appear as relatively

---

[1] Specific-resistance values between 10^7 and 10^{15} Ω cm at 300 K are reported in the literature for the transition-metal monoxides [1, 27, 33, 127], [3, p.5]. The specific resistance can differ by several orders of magnitude for different samples of the same compound. This must be attributed to impurities and defects, which are always present in these oxide crystals and determine the conductivity [1] (Sect. 4.1.4).

sharp structures in electron energy-loss spectra [56, 63, 64, 184] (Figs. 5.7 and 5.32), similar to excitations of free atoms. Strong final-state effects, due to the electron correlation, are observed in valence-band as well as core-level photoemission spectra through the existence of strong satellites [25, 59, 84, 97, 105, 113, 114, 124, 178, 179, 200, 201, 202, 203, 211]. In particular, the sharp structures observed in the optical gap in optical absorption and electron energy-loss spectra can only be interpreted as transitions between localized, discrete transition-metal 3d states, which are split in the crystal field of the surrounding oxygen ions (for a survey see the book by Cox [23]; several other publications are cited in Sect. 2.3).

It is well accepted now that local models, such as crystal-field, ligand-field, or cluster calculations, are needed for realistic theoretical treatments of the transition-metal 3d states (Sect. 2.3) which are consistent with experimental results. For the O 2p states, on the other hand, LDA band-structure calculations have been found to be in fairly good agreement with experimental results, as a comparison with dispersion curves deduced from angle-resolved photoemission spectra of NiO and CoO shows [82, 83, 177, 178, 179], [84, p. 276]. It is this coexistence of local and band-like features in the electronic structure and the resulting physical properties such as the occurrence of the large insulating gaps in the transition-metal monoxides which led to the long-standing interest in these materials, resulting in numerous investigations. In particular, in the last few years, the attempts to understand the electronic structure and high-temperature superconductivity of the Cu–O perovskites led to a reexamination of the more or less unsolved problems occurring in the similar but much simpler binary transition-metal oxides and forced the intensification of theoretical as well as experimental investigation of the electronic structure of these compounds. But at present, no unified theoretical approach is available which can describe the electronic properties of the monoxides coherently [23, p. 36], [84, p. 183].

In this chapter, the electronic structure of NiO, CoO, and MnO is briefly reviewed. Apart from a short summary of the origin of the insulating gaps (Sect. 2.2), the main considerations are the localized 3d states, their crystal-field splitting, and the d–d excitations within this crystal-field multiplet (Sect. 2.3), because these states and excitations have been the main subject of our experimental investigations.

2.2 Crystal Structure and Optical Gap

The transition-metal monoxides MnO, CoO, and NiO form ionic, anti-ferromagnetic crystals with the NaCl structure [3, p. 4] (Fig. 2.1). Two transition-metal electrons saturate the O 2p shell, leading to O^{2-} ions with the $[He]2s^2 2p^6$ configuration and TM^{2+} ions with the $[Ar]3d^n$ configuration (Table 2.1).

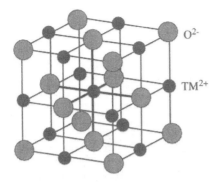

Fig. 2.1. Crystal structure of MnO, CoO, and NiO

Table 2.1. Electronic configuration, measured insulating gap, Néel temperature, and lattice constant of MnO, CoO, and NiO. The data are taken from the references indicated in the footnote

Oxide	Electronic Configuration		Insulating gap (eV)	Néel Temperature (K)	Lattice Constant (nm)
	O^{2-}	TM^{2+}			
MnO	[He] $2s^2 2p^6$	[Ar] $3d^5$	3.6–4.2[a]	118[d]	0.444[d]
CoO	[He] $2s^2 2p^6$	[Ar] $3d^7$	2.5–6[b]	289[d]	0.426[d]
NiO	[He] $2s^2 2p^6$	[Ar] $3d^8$	3.1–4.3[c]	523[d]	0.417[d]

[a][21, 33, 94, 201].
[b][163, 178, 202], [64, p. 72], [21, 164].
[c][63, 82, 83, 108, 127, 163, 170].
[d][3, p. 5].

The chemical bond is not purely ionic, but also contains covalent contributions, as can be directly inferred from the results of x-ray absorption spectroscopy (XAS). Without any hybridization between anion and cation orbitals, the dipole-allowed O 1s–O 2p transition is impossible owing to the closed O 2p shell of the O^{2-} ion. Nevertheless, the x-ray absorption spectra exhibit a high intensity near the O 1s threshold [26, 28, 105, 108, 141], which is attributed to the reduction of the number of filled states with O 2p character owing to O 2p/TM 3d hybridization. The strength of the O 1s-O 2p/TM 3d absorption is directly related to the degree of covalency, which is strong in the first half of the 3d series and diminishes for the late transition-metal oxides owing to shrinkage of the TM 3d orbitals [28], [84, p. 202]. A hybridization of O 2p with TM 4s and TM 4p states is also found [28]. The O 2p–TM 3d hybridization of MnO and NiO is also evident from results of resonant photoemission spectroscopy at the Mn 2p threshold [200, p. 77] and from comparison of x-ray emission spectra (O 2p–O 1s and Ni 3d–Ni 2p transitions) [200, p. 54].

The antiferromagnetic order of the transition-metal monoxides MnO, FeO, CoO, and NiO is of that type which is often called type II antiferromagnetism [66, pp. 91ff.], [152, 195, 199]: in the transition-metal fcc sublattice (Fig. 2.1), nearest neighbors in the [100] direction (which are separated by an oxygen ion) are antiferromagnetically coupled, i.e. their spins are in opposite directions. As result, the (111) planes of the transition-metal sublattice are planes of parallel spins; adjacent (111) planes show an antiparallel alignment of the spins. The Néel temperatures, which differ strongly for the different oxides, are given in Table 2.1.

From a simple band-structure point of view, NiO, CoO, and MnO should be metals like the corresponding transition metals, owing to their incompletely filled 3d states. But, as mentioned already in Sect. 2.1, they are insulators with wide insulating gaps[2] of several electron volts. The gap widths have been determined by several experimental methods, such as optical absorption spectroscopy [164], [94, 163], electron energy-loss spectroscopy [63], [64, p. 72], and photoconductivity [33] and electroreflectance [127] measurements. In addition, the gap widths can be inferred from the energy separation between occupied and unoccupied states by combining photoemission and bremsstrahlung isochromat spectra [82, 83, 170, 178, 201, 202]. It is also possible to estimate the gap width of NiO by comparison of x-ray absorption spectra of pure and Li-doped NiO [83, 108].[3] The results obtained by the different methods are summarized in Table 2.1. The differences in the published gap widths arise mainly from different gap definitions, and it seems to be more or less a matter of taste which is preferred. For NiO, for example, the values obtained by different methods of evaluation from one single absorption curve deviate by more than 25%, as illustrated by Hüfner [83], [84, p. 188]. In the case of CoO this deviation is considerably larger (Sect. 5.6.2), which is the reason for the larger scatter of the published gap widths (Table 2.1). Some of the gap definitions which can be used to obtain the gap width from optical absorption and electron energy-loss spectra are described in Sect. 5.6.2, where the optical gap widths are determined from our electron energy-loss spectra (Fig. 5.33).

---

[2] The term "insulating gap" or "optical gap" instead of "bandgap" is used here to emphasize the fact that the origin of the gaps in NiO, CoO, and MnO is not describable in terms of single-particle band-structure calculations. The electronic transitions which determine the gap width do not occur between band-like states, but involve localized d states as explained below.

[3] The first peak in the x-ray absorption spectra of the NiO oxygen K edge (532 eV) is attributed to an $O\,1s^2\,Ni\,3d^8$–$O\,1s\,Ni\,3d^9$ transition, which is possible owing to the hybridization of the O 2p and Ni 3d states (see above). In the Li-doped samples a further absorption peak appears at ≈ 528.5 eV, which is assigned to the excitation of O 1s electrons into holes created by the doping. The energy difference between these two peaks is close to but smaller than the gap width, because the energy separation between the Fermi level and the Li induced hole must be taken into account.

The origin of the gaps has been the subject of numerous investigations since the discovery of the insulating nature of the transition-metal monoxides. These investigations and their results have been reviewed by several authors. The present "state of the art" is given in a detailed review by Hüfner [84] and the references therein and is summarized here only briefly.

As sketched already in Sect. 2.1, the difficulties in understanding the nature of the insulating gaps and many other physical properties of the transition-metal monoxides arise from the fact that no unified theory exists which covers all the properties of the oxides sufficiently. Whereas the fully occupied O 2p states form bands with a dispersion consistent with LDA band-structure calculations, as shown by comparison with experimental results obtained by angle-resolved photoemission spectroscopy [82,83,177,178,179], the application of such calculations to the 3d states produces results in contradiction to many experimental results: from the angular-resolved photoemission measurements of Shen et al. [177,178], the dispersion of the 3d states is found to be $\approx 0.3\,\mathrm{eV}$ for NiO(100) and CoO(100) single crystals, which is only 25% of the 3d band width (1.2 eV) calculated by the same authors. A similar small dispersion is also measured by Hüfner et al. [82, 83] and Kuhlenbeck et al. [105] for cleaved NiO(100) single crystals, as well as NiO(100)/Ni(100) thin films. Also, the dispersion of the MnO 3d states is small and found to be less than $\pm 0.1\,\mathrm{eV}$ [113]. The agreement between experimental results and band-structure calculations is improved if the antiferromagnetic order is taken into account. In this case, the calculated 3d bands of NiO are found to be much narrower than the nonmagnetic bands [179], but several discrepancies still remain. In particular, single-particle band-structure calculations cannot reproduce the measured gap widths of the oxides: in such calculations, small insulating gaps occur for MnO and NiO, if type II antiferromagnetism is assumed [151, 152, 194, 195]. In the paramagnetic phase a very small gap persists for MnO, but in the case of NiO the gap vanishes completely. This would lead to a conducting behavior of NiO above the Néel temperature, which is not observed experimentally. The gap widths of $0.4\text{--}2.2\,\mathrm{eV}$ for MnO [12, 104, 151, 152, 190, 194, 195] and $0.2\text{--}0.4\,\mathrm{eV}$ for NiO [12, 87, 151, 152, 190, 194, 195] are too low and deviate for NiO, in particular, by an order of magnitude from the experimental results (Table 2.1). Independently of the magnetic order, no gap is found for CoO at all: CoO should be a conductor from single-particle band-structure calculations, in contradiction to all experimental results [12, 194, 195]. The reason for these discrepancies is the strong Coulomb interaction among the 3d electrons in the oxides, which cannot be treated correctly in single-particle band-structure calculations. This Coulomb correlation prevents the electrons from forming 3d bands and localizes them at the transition-metal ions.

It is well established now that the insulating gaps of the transition-metal monoxides do not occur between band-like states. "Interionic" excitations between two TM^{2+} ions with transfer of a 3d electron from one transition-metal ion to another determine the gap width. It is customary to distinguish between two kinds of such interionic transitions, the *Mott–Hubbard* and the *charge-transfer* transitions. The two transitions usually require different excitation energies, owing to different final states. These two kinds of interionic transitions are sketched in Fig. 2.2: for the excitation of the Mott–Hubbard transition (Fig. 2.2a), the Coulomb correlation energy U is needed for the creation of a 3d hole at one transition-metal site and transfer of the electron to another transition-metal ion:

$$3d^n + 3d^n + U \rightarrow 3d^{n-1} + 3d^{n+1} \tag{2.1}$$

If the 3d hole is immediately screened by a charge transfer from the oxygen ligand (Fig. 2.2b), the hole is finally located at the ligand and the charge-transfer energy Δ is required for this excitation:

$$3d^n L + 3d^n L + \Delta \rightarrow 3d^n L^{-1} + 3d^{n+1} L \tag{2.2}$$

In (2.2), the oxygen ligand is denoted by L and the ligand hole, due to the missing electron by L^{-1}.

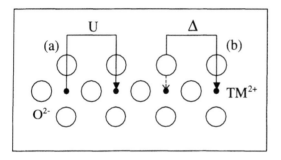

Fig. 2.2. Interionic transitions between TM^{2+}-ions (according to Zaanen and Sawatzky [171, 215]): (**a**) Mott–Hubbard transition; (**b**) charge-transfer transition

If $U < \Delta$, the Mott–Hubbard transition determines the gap and the compound is called Mott–Hubbard insulator; if $\Delta < U$, it is called a charge-transfer insulator. Compounds where U and Δ are of comparable size are a mixed form of the two types of insulator because both Mott–Hubbard and charge-transfer transitions contribute to the excitations which determine the gap.

NiO was thought to be the prototype of the Mott–Hubbard insulator for a long time, but it is clear now that it belongs to the class of charge-transfer insulators. Values for U and Δ are obtained by comparing and fitting the results of cluster calculations (see Sect. 2.3.1) to core level or valence band[4]

---

[4] The term "Valence band" is chosen here according to the customary usage in photoemission spectroscopy (PES). In PES, emission from the occupied states

photoemission spectra [58, 114, 203]. Many of these values have been summarized by Hüfner [84, p. 226]; values obtained from other calculations and measurements are also included. In most of the calculations Δ is found to lie between ≈ 4–$6\,\text{eV}$ and is therefore several electron volts lower than U, which is calculated to be ≈ 7–$10.5\,\text{eV}$. In our electron energy-loss measurements we find the maximum of the first excitation across the optical gap at ≈ 4.8–$5\,\text{eV}$ (shoulder a in Fig. 5.33a, Sect. 5.6.2), which is in accordance with the values calculated for the charge-transfer transition.

MnO seems to be a mixed form of Mott–Hubbard and charge-transfer insulator. The Coulomb correlation energy and the charge-transfer energy have been calculated to be of comparable size ($\Delta = (7 \pm 1)\,\text{eV}$, $U = (7.5 \pm 0.05)\,\text{eV}$ [60]; $\Delta = 8.8\,\text{eV}$, $U = 8.5\,\text{eV}$ [202]). Our measurements are in accordance with these values (Sect. 5.6). The maximum of the first excitation across the gap occurs at $\approx 7\,\text{eV}$ (Table 5.5, shoulder a in Figs. 5.32 and 5.33c).

For CoO, the situation is not yet clear [114, 202]. Whereas most of the calculated values of the charge-transfer energy Δ are between 5 and $7\,\text{eV}$, the values of the Coulomb correlation energy U scatter considerably (4.7–11 eV) [84, p. 214]. The gap of CoO seems to be of either the charge-transfer or the mixed charge-transfer/Mott–Hubbard type with $\Delta \approx U$; both assumptions are in accordance with the published values for U and Δ. Optical absorption measurements exhibit a first absorption peak at $\approx 6.5\,\text{eV}$ [163]. A weak excitation shoulder at 6.5–$7\,\text{eV}$ is also indicated in electron energy-loss spectra [65], [64, pp. 71ff] (shoulder a in Fig. 5.33b), but a broad, distinct maximum occurs between 7.5 and $9.5\,\text{eV}$ (b in Fig. 5.33b; $\approx 8.7\,\text{eV}$ in the spectra of Gorschlüter and Merz [65], [64, pp. 71ff]). Both measured excitation energies roughly coincide with the published values for Δ and U. In our energy-loss spectra an equivalent behavior of the gap transitions of MnO and CoO is observed, unlike that of NiO (Sect. 5.6.2). This may be indicative of an identical gap type in CoO and MnO: a mixed Mott–Hubbard/charge-transfer gap must also be assumed for CoO.

Recent theoretical work on the transition-metal oxides introduces more realistic electron interactions and correlations into band-structure calculations and therefore tries to integrate the Mott–Hubbard or charge-transfer picture into band models. Here, the calculated gap widths of 2.54–4.5 eV for NiO [5, 118, 190, 191, 192, 193] and 3.4–5 eV for MnO [5, 104, 120, 190, 191, 193] are in good agreement with the measured values (Table 2.1). Also, for CoO, realistic gap values of 2.51–4 eV are obtained with such calculations [5, 190, 191, 193]. In the calculations for NiO of Manghi et al. [118], for example, no assumption about magnetic order was made – the calculated gap is present even in the paramagnetic phase, and NiO remains insulating above the Néel temperature. In addition, in these calculations, the main structures of photoemission

---

close to the Fermi level is called valence-band emission even in the transition-metal oxides, despite the fact that band-like O 2p as well as localized TM 3d states are involved.

spectra of the valence band as well as the inverse photoemission spectra and O K$_\alpha$ x-ray emission spectra are often well reproduced [118,120,192,193]. But the search for a unified theory remains: like all other band-structure calculations, these models are also unable to explain the existence of the crystal-field multiplet arising from the splitting of the 3d states in the crystal field of the surrounding O^{2-} ions. The crystal-field splitting and the d–d excitations observed within this crystal-field multiplet can be understood in the framework of local models only, which are discussed in the following section.

2.3 3d States and d–d Transitions

2.3.1 Crystal-Field Multiplets

As pointed out above (Sect. 2.2), it is the strong Coulomb correlation of the electrons in the incompletely filled TM 3d shell which is responsible for the appearance of the large insulating gaps in the transition-metal monoxides. This correlation localizes the 3d electrons at the TM^{2+} ions. The 3d states are not describable in a pure band picture, and the 3d electrons behave similarly to the d electrons of free atoms or ions, but with one significant difference: in contrast to free atoms or ions, the degeneracy of the 3d states concerning the magnetic quantum number m_ℓ is partially lifted. In the O_h-symmetric crystal field, provided by the six O^{2-} ions surrounding each transition-metal ion octahedrally (Fig. 2.1), the 3d states, which are completely degenerate in the spherical symmetry of a free atom, are energetically split and a *crystal-field multiplet* of 3d states occurs. The crystal-field splitting is often illustrated and explained in a "one-electron picture", where *single d electrons* are considered in an O_h-symmetric environment of *point charges*, representing the oxygen ions (Fig. 2.3).

Whereas the e$_g$ orbitals (d$_{z^2}$ and d$_{x^2-y^2}$) are directed towards the oxygen ions, located on the coordinate axes, the t$_{2g}$ orbitals (d$_{xy}$, d$_{xz}$, and d$_{yz}$) are located between the oxygen ions. Therefore, a 3d electron in an e$_g$ orbital experiences a Coulomb potential different from that acting on a 3d electron in a t$_{2g}$ orbital, and the t$_{2g}$ and e$_g$ states become energetically separated by Δ_{CF} (Fig. 2.3). Δ_{CF} is called the "crystal-field splitting" or "crystal-field splitting parameter" and is the energy required for the excitation of a single d electron in the crystal field from a t$_{2g}$ into an e$_g$ orbital. In terms of group theory, the eigenfunctions corresponding to the t$_{2g}$ and e$_g$ states of single d electrons belong to different irreducible representations of the O_h point group and have different eigenvalues [22].

If the one-electron picture is extended by introducing the spin of the electrons and the exchange splitting Δ_{Ex}, which occurs between t$_{2g}$ or e$_g$ states of opposite spin directions (Fig. 2.4), this simple model can also be used to understand the electron configuration of the ground state and its multiplicity: in MnO (Fig. 2.4a), the exchange splitting Δ_{Ex} exceeds the crystal-field

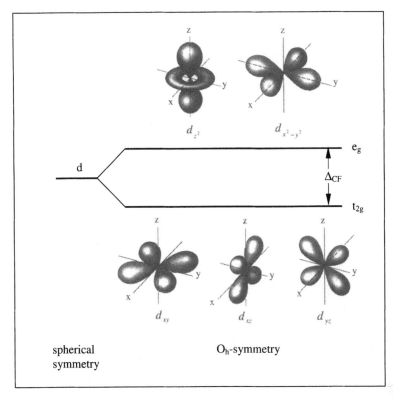

spherical
symmetry

O_h-symmetry

Fig. 2.3. Splitting of the d states of single electrons in an O_h-symmetric crystal field (after [23, p. 38])

splitting Δ_{CF} considerably [180, 195]. The ground state of the Mn^{2+}-ions in MnO therefore has a high-spin configuration, i.e. the five d electrons occupy states of parallel spin preferentially, leading to a $t_{2g}^3 e_g^2$ configuration and a sextet ground state. Progressing through the series of monoxides, the exchange splitting decreases with increasing number of d electrons [194]. Nevertheless, $\Delta_{Ex} > \Delta_{CF}$ for CoO. The ground state of the Co^{2+} ions in CoO then also has a high-spin configuration, and Fig. 2.4a can be applied to CoO, with its $3d^7$ configuration (Table 2.1), by adding two $t_{2g}(\downarrow)$ electrons (not shown in Fig. 2.4a). This leads to a $t_{2g}^5 e_g^2$ configuration and a quartet ground state [23, pp. 39, 40]. For NiO, the situation is slightly different, because the crystal-field and exchange splittings are of comparable size ($\approx 1\,\mathrm{eV}$) [195]. However, even in the case that Δ_{CF} exceeds Δ_{Ex} slightly, the Ni^{2+} ions have a high-spin ground state owing to their $3d^8$ configuration, as can be inferred from Fig. 2.4b. All t_{2g} levels and the two e_g levels of parallel spin are occupied ($t_{2g}^6 e_g^2$ configuration) and NiO has a triplet ground state [23, pp. 39, 40].

For the determination of a realistic crystal-field splitting Δ_{CF}, the one-electron model is oversimplified, because the assumption of a pure Coulomb

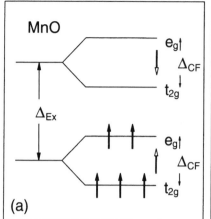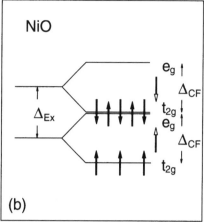

Fig. 2.4. Crystal-field splitting Δ_{CF} and exchange splitting Δ_{Ex}: (a) MnO; (b) NiO. (a) also applies to CoO, if two $t_{2g}(\downarrow)$ electrons are added. In MnO, all d states of parallel spin are occupied, whereas those of opposite spin are completely empty. Similar figures have been published by several authors [65, 171, 202]

interaction between TM 3d electrons and O 2p ions usually underestimates the size of the crystal-field splitting. Orbital overlap and covalent interactions between metal and oxygen electrons must be taken into account (for a survey see the book by Cox [23, pp. 36ff.] or the review by Brandow [15] and references therein), and often these interactions give the main contributions to the crystal-field splitting: in NiO, Δ_{CF} is found to be ≈ 1.1 eV [44, 148], but calculations of the crystal-field splitting based on the Coulomb interaction of the d electrons with point charges representing the oxygen ions give a much smaller value of only 0.3 eV [44].

Although the simple one-electron picture provides the great advantage of intuitive understanding of the occurrence of the crystal-field splitting (but not its size) and the occupation of the different d orbitals in the ground state, it must be used carefully and one has to be aware of its limitations: for the understanding of the variety of excited crystal-field multiplet states, the one-electron picture is often misleading, because not single d electrons but the Russell–Saunders terms of the complete transition-metal ion, arising from the LS coupling of its five, seven, or eight d electrons (Table 2.1) must be considered in the crystal field. Owing to the identical dependence of the single-electron wave functions and the wave functions corresponding to the Russell–Saunders terms S, P, D, F... of the complete d^n configuration on the magnetic quantum number m_ℓ, the Russell–Saunders terms are split in the crystal field in the same manner as the single-electron states with the corresponding angular momenta, s, p, d, f... [22, p. 253]: as illustrated in Fig. 2.3 for the d orbitals of single electrons, the ^{1}D state of the eight 3d electrons of a free Ni^{2+} ion, for example, splits into a threefold degenerate

Table 2.2. Possible Russell–Saunders terms of the d^5, d^7, and d^8 configurations of free ions ([4, p. 166])

Configuration	Terms
d^5 (Mn^{2+})	^6S, ^4G, ^4F, ^4D, ^4P, ^2I, ^2H, ^2G, ^2F, ^2D, ^2P, ^2S
d^7 (Co^{2+})	^4F, ^4P, ^2H, ^2G, ^2F, ^2D, ^2P
d^8 (Ni^{2+})	^3F, ^3P, ^1G, ^1D, ^1S

T_{2g} and a twofold degenerate E_g level (now also denoted by capital letters) if the Ni^{2+} ion is put into the O_h-symmetric crystal field of NiO. All possible Russell–Saunders terms of the d^5, d^7 and d^8 configurations and their splitting in the O_h-symmetric crystal field are listed in Table 2.2 and 2.3.

More instructive than Tables 2.2 and 2.3 are the term schemes of the 3d crystal-field multiplets. Such term schemes are plotted in Figs. 2.5–2.7 for MnO, CoO, and NiO in anticipation of our experimental results and the discussion below. The term energies in Figs. 2.5–2.7 correspond to values obtained by us with scattering-geometry-dependent spin-polarized electron energy-loss spectroscopy (Sect. 5.2, 5.3, and 5.5, Tables 5.1–5.3), with a few exceptions only. The assignment of the measured energies to the different crystal-field multiplet-terms has been done according to the results of our measurements in comparison with calculations and other experimental results; details are discussed in Sect. 5.5.

From the application of group theory, the number and symmetry of the possible crystal-field multiplet terms is known for the different d^n configurations in a crystal field of a given symmetry (Tables 2.2 and 2.3 for O_h symmetry; see the book by Cotton ([22, pp. 252, 253] for other symmetries), but nothing can be inferred about their actual energies (Figs. 2.5–2.7). These energies depend strongly on the crystal field and on the electrostatic repulsion between the d electrons. Spin–orbit interaction is small in the oxides of the 3d series and often neglected in calculations [23, p. 43], [183]. The crystal-field strength is determined by the interactions between the transition-metal and

Table 2.3. Splitting of the Russell–Saunders terms in an O_h-symmetric crystal field ([22, pp. 252ff.], [188, p. 33])

Term	Splitting in an O_h-symmetric Crystal field
S	A_1
P	T_1
D	$E + T_2$
F	$A_2 + T_1 + T_2$
G	$A_1 + E + T_1 + T_2$
H	$E + 2T_1 + T_2$
I	$A_1 + A_2 + E + T_1 + 2T_2$

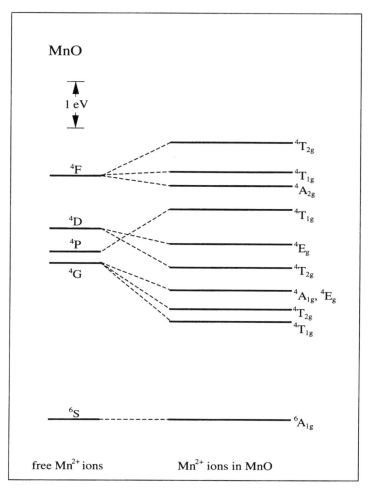

Fig. 2.5. Crystal-field splitting of MnO. All sextet and quartet states of the crystal-field multiplet in O_h symmetry are shown; the doublet states (Table 2.3) are omitted. The energetic positions correspond to our experimental results, except for the highest $^4T_{2g}(^4F)$ state (Sect. 5.5.2, Table 5.1). The energies of the Russell–Saunders terms of the free ions are taken from [188, p. 111]

oxygen ions and their electrons. As pointed out above, orbital overlap and covalent contributions play a dominant role here and exceed the contribution of the pure Coulomb interaction. The crystal-field strength depends strongly on geometric parameters such as bond lengths or lattice constants. This dependence is the reason why ruby (and probably other TM oxides) can be used for measurements of very high pressure ($> 10^9$ Pa), as already mentioned in the introduction (Chap. 1): ruby consists of sapphire (Al_2O_3) with less than 1% of the aluminum substituted by chromium. Owing to the compression of the crystal under high pressure, the crystal-field strength changes. Thus, the

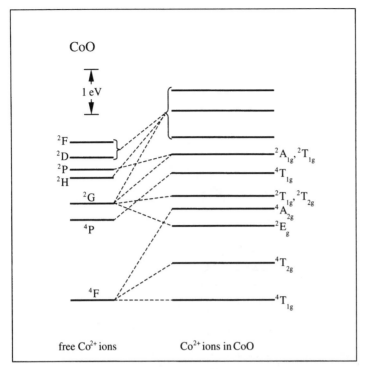

Fig. 2.6. Crystal-field splitting of CoO. All quartet states of the crystal-field multiplet in O_h-symmetry are given; some of the doublet states of higher energy (Table 2.3) are omitted. The energetic positions of the energy levels correspond to our experimental results, except for the ^2E_g (^2G) level (Sect. 5.5.3, Table 5.2). Most energies of the Russell–Saunders terms of the free ions are taken from ([188, p. 109]); some of the higher doublet terms (^2P, ^2D, ^2F) are arbitrarily chosen

energy separations within the crystal-field multiplet of the Cr^{3+} ions change, and the energy of the light emitted or absorbed in d–d transitions is shifted under pressure. The energy of the strongly dipole-forbidden (Sect. 2.3.2) $^4A_{2g}$ $\rightarrow$ ^2E_g transition, used for high-pressure measurements, decreases slightly but linearly with pressure (≈ 0.8–$0.9\,cm^{-1}$ per $10^8\,Pa$) [42, 155].

Several calculations of the crystal-field multiplets of transition-metal compounds have been performed in the last forty years, differing mainly in the treatment of the orbital overlap and mixing between the TM 3d and O 2p states and the repulsion between the d electrons. A brief survey of the different theoretical models is given in the book by Cox [23, pp. 36ff.]: in the earlier ligand-field calculations, leading to the Sugano–Tanabe and Orgel diagrams [188, pp. 106ff.], [22, pp. 265ff.], [154], it is assumed that the d electrons remain in atomic-like d orbitals in the crystal field, and the electron repulsion obtained for free ions with the corresponding d^n configuration is used. The energies of the 3d multiplet terms of ions in different crystals are plot-

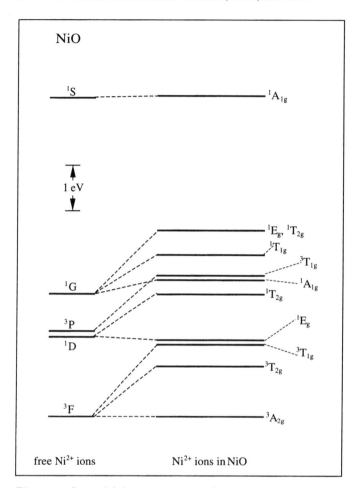

Fig. 2.7. Crystal-field splitting of NiO. All components of the crystal-field multiplet in O_h symmetry (Table 2.3) are given. The energetic positions mainly correspond to our experimental results, some calculated singlet terms are also included (Sect. 5.5.4, Table 5.3); the separation of very close-lying levels has been slightly enhanced. The energetic separations between the Russell–Saunders terms of the free ions are taken from ([188, p. 108])

ted as a function of the crystal-field strength, represented by the splitting parameter Δ_{CF} (Figs. 2.3 and 2.4). Owing to a reduction of the electron repulsion in the transition-metal compounds [23, pp. 43,44] and uncertainties in the actual value of Δ_{CF} for the compounds considered (see Sect. 5.5.2), the Sugano–Tanabe diagrams often give only a rough quantitative estimation of the energy levels. However, they are very helpful for determining the crystal-field multiplet of a compound qualitatively, and the energies taken from these diagrams often show an astonishingly good correspondence with experimental results (Sect. 5.5.3, Table 5.2). The diagrams can be used in a very successful way to understand the differences in the multiplets of identical ions in different crystal-field surroundings, thereby explaining the different colors of ruby and Cr_2O_3, for example. In these two compounds, different crystal-field strengths are responsible for different energetic separations of the multiplet states of Cr^{3+} ions in Al_2O_3 and Cr_2O_3 [188, pp. 117ff.], [125]. Light of different wavelengths is absorbed in the d–d excitations within the crystal-field multiplets, leading to the red and green colors, respectively, of these substances.

Especially in the last few years, more sophisticated local *cluster models*, based on molecular-orbital and configuration-interaction calculations, have been developed and applied very successfully in calculations of the 3d crystal-field multiplets of several transition-metal compounds [13,44,59,60,73, 181, 201, 202, 203]. In these calculations small clusters, often containing the transition-metal ion and its adjacent six oxygen ions only, are considered. Apart from calculations of the multiplet of the bulk transition-metal ions, the efficiency of such cluster calculations was impressively demonstrated in calculations of the crystal-field multiplet of the surface transition-metal ions, which deviates from that of the bulk owing to changes in the crystal field caused by missing oxygen ions (Sect. 2.3.3.1). The calculated energies of the surface crystal-field multiplet [13, 44, 128, 181] are in fairly good accordance with experimental results [13, 44, 53, 65, 73, 139, 140] (Sects. 5.3.2, 5.5.3, 5.5.4, Table 5.4).

Cluster calculations are also found to be an effective tool for the interpretation and explanation of x-ray absorption, bremsstrahlung isochromat and, in particular, photoemission spectra of transition-metal compounds [58, 59, 60, 113, 114, 153, 201, 202, 203, 214]. The photoemission spectra exhibit a variety of satellite structures, arising from the strong electron correlation. With the configuration-interaction method the contributions of unscreened $d^{n-1}L$ final states, where the photoelectron is emitted from a transition-metal 3d state, and $d^n L^{-1}$ final states, where the 3d hole is screened by a ligand-to-metal charge transfer (L^{-1} denotes the ligand hole; see Sect. 2.2), to the main lines and the various satellites could definitely be determined. The participation of $d^{n+1}L^{-2}$ final states with transfer of two electrons from the ligand to the metal has also been demonstrated for NiO [58, 59, 203], as well as for MnO [60, 201] and CoO [202].

All theories, ligand-field as well as cluster calculations, agree in the predictions for the ground states of the TM^{2+} ions in MnO, CoO, and NiO: the ground states of free Mn^{2+}, Co^{2+} and Ni^{2+} ions are formed according to Hund's rules. The state with the highest multiplicity has the lowest energy. If more than one state exists with this multiplicity, the one with the highest angular momentum is the ground state. This is the ^6S state for Mn^{2+} ions, ^4F for Co^{2+} ions, and ^3F for Ni^{2+} ions (Table 2.2). Owing to their high-spin configurations (Fig. 2.4), the ground state of a transition-metal ion in the crystal field of the corresponding monoxide is the lowest crystal-field component of these states (Table 2.3). These are the $^6A_{1g}$, $^4T_{1g}$, and $^3A_{2g}$ states in MnO, CoO, and NiO [154], [188, pp. 106ff.]. Whereas the ^4F and ^3F ground states are split into three states of A_{2g}, T_{1g}, and T_{2g} symmetry in the O_h-symmetric crystal field of CoO and NiO (Table 2.3, Figs. 2.6 and 2.7), the high-symmetry ^6S state of Mn^{2+} ($^6A_{1g}$ for Mn^{2+} ions in MnO) remains unsplit. All excited states of free Mn^{2+} ions, as well as those of Mn^{2+} ions in MnO, are quartet or doublet states (Tables 2.2 and 2.3, Fig. 2.5). In contrast to CoO and NiO, all d–d transitions from the ground state therefore require a change in multiplicity. The d–d transitions in NiO, CoO, and MnO and their probabilities are discussed in Sect. 2.3.2.

A variety of measured and calculated term energies of the excited 3d crystal-field multiplet states have been published (see Tables 5.1–5.3 and references cited above). Most calculations have been carried out and optimized in relation to experimental results and therefore often show excellent correspondence with some of the measured multiplet terms. Nevertheless, a lot of open questions remain. Owing to particularly strong discrepancies between different calculations and the nearly complete lack of calculations and measurements of most of the energies of the higher multiplet terms, a variety of uncertainties arise in the assignment and comparison of the measured energies to the calculated values. The details are discussed in Sect. 5.5 together with our experimental results, and only a few examples will be given here: for CoO (Fig. 2.6), the structure of the crystal-field multiplet seems not to be clear at all at present. Different calculations differ significantly for several 3d states, and the assignment of measured energies to particular multiplet terms is therefore very difficult. Even the energy of the $^4A_{2g}$ state, split from the ^4F ground state of the free d^7 configuration, differs by more than 1 eV in different cluster calculations (the calculated values are 1.71 eV [73, 181] and 3.06 eV [202], Table 5.2), and none of these calculations is very close to the experimental results, which clearly show an energy of 2 eV for the $^4A_{2g}$ (^4F) state (Sects. 5.3.1.1, 5.5.3). For NiO (Fig. 2.7), the $^3T_{1g}$ (^3F) and ^1E_g (^1D) states are calculated to differ by $\approx$ 50–300 meV only, but different calculations make different predictions as to whether the $^3T_{1g}$ or the ^1E_g state is higher in energy [59, 96, 129, 203] (Table 5.3).

2.3.2 d–d Transitions

One thing in particular that is responsible for the present uncertainties in the crystal-field multiplets is the lack of definite experimental data for a variety of term energies, especially for those of higher excitation energy. This lack is caused by the considerable experimental difficulties arising in the measurement of these energies: excitations within the crystal-field multiplet are transitions between 3d states and therefore strongly forbidden by dipole selection rules. All d–d transitions violate the parity selection rule $\Delta\ell = \pm 1$ (Laporte rule). The multiplicity-changing transitions violate the spin selection rule $\Delta S = 0$ additionally. Whereas in NiO and CoO multiplicity-conserving triplet–triplet and quartet–quartet transitions, as well as multiplicity-changing triplet–singlet and quartet–doublet transitions, occur (Figs. 2.7 and 2.6), in MnO, with its half-filled 3d shell (Table 2.1) all transitions from the $^6A_{1g}$ (^6S) ground state are multiplicity-changing (Fig. 2.5) and are forbidden by the parity as well as by the spin selection rule.

Generally, electric dipole-forbidden transitions remain forbidden in O_h symmetry, but the parity selection rule is slightly weakened by a mechanism initially introduced and discussed by van Vleck for the f–f transitions of rare earths [206]. Briefly, an admixture of odd-parity parts to the even-parity wave function can occur even in complexes with a center of inversion, owing to symmetry distortions arising from lattice vibrations [188, p. 113], [22, pp. 280ff.], [23, p. 46]. Nevertheless, the dipole matrix elements remain small, leading to optical absorption coefficients two to three orders of magnitude lower than those of dipole-allowed transitions across the optical gaps of the oxides [148, 163, 164]. Those transitions which are forbidden additionally by the spin selection rule can occur in optical absorption spectra only because of spin–orbit interaction [188, p. 116], which is expected to be small in the oxides of the 3d series [23, p. 43]. The optical transition probabilities for parity- *and* spin-forbidden electric dipole transitions have been estimated to be seven orders of magnitude lower than for the dipole-allowed ones [188, p. 116]. Nevertheless, the spin-forbidden transitions are visible in optical absorption spectra, but with very weak intensities [85, 94, 148, 164].

More convenient for the examination of d–d transitions is electron energy-loss spectroscopy (EELS) with low-energy electrons. Especially in the last few years, this method has gained increasing significance in investigations of the dipole-forbidden d–d excitations in transition-metal oxides and f–f excitations in rare-earth metals (see Sect. 3.1 for references). The great advantage of exciting such transitions with slow electrons is the possibility of *excitation by electron exchange*. Multiplicity-conserving ($\Delta S = 0$), as well as multiplicity-changing transitions with $\Delta S = -1$, are easily observable with electron energy-loss spectroscopy if a suitable energy of the incident electrons is chosen (Sects. 5.1, 5.2). In particular, if spin-polarized electron energy-loss spectroscopy (SPEELS) is used, where the d–d excitations are excited by polarized electrons and the polarization of the scattered electrons is measured (Sect. 3.3), the exchange processes can be proved unambiguously and their contributions to the excitation process can be determined for the different d–d transitions. This is discussed in detail in Chap. 3. Other advantages of electron energy-loss spectroscopy in comparison with optical absorption spectroscopy are also discussed there.

2.3.3 Surface Transition-Metal Ions

2.3.3.1 Surface Crystal-Field Multiplets. As already pointed out in Sect. 2.3.1, the 3d crystal-field splitting of transition-metal ions in compounds depends on both the strength and the symmetry of the crystal field. It is intuitively clear, then, that the crystal-field multiplet of surface TM ions in oxides differs strongly from that of the bulk ions, owing to the reduced symmetry caused by missing oxygen ions at the surface (Fig. 2.8). In fact, in the electron energy-loss spectra of NiO, CoO, and Cr_2O_3 sharp energy-loss peaks appear in the gap which cannot be attributed to excitations within the bulk 3d multiplet (Sect. 2.3.1). These peaks are visible only with the very surface-sensitive version of EELS that uses low-energy electrons (Sect. 3.1) – not in spectra obtained with surface-insensitive methods such as high-energy EELS [63, 65], [64, pp. 103ff.] and optical absorption spectroscopy [148] – and they are found to depend on the surface and its properties: they appear in spectra of freshly cleaved transition-metal oxide crystals or freshly prepared oxide films only and are very sensitive to adsorbates, contamination, and surface damage by electron or ion impact [17, 44, 65, 106, 213], [64, pp. 103ff.], [13, 46, 47, 48, 53, 54, 73, 107]. In electron energy-loss spectra of very thin NiO/Ag(001) films (in the submonolayer range) only these peaks are visible in addition to the flat EEL spectrum of the Ag substrate – bulk d–d excitations are not observed as expected [139, 140].

With cluster calculations [13, 44, 73, 128, 181], it was possible to assign these energy-loss structures to excitations within the 3d crystal-field multiplet of the surface transition-metal ions. In these calculations, the bulk TM-ions of oxides with the NaCl structure such as NiO and CoO (Sect. 2.2) are considered in the octahedral crystal field of the surrounding six O^{2-} ions (Fig. 2.8), embedded in an infinite Madelung field of point charges, representing the other ions of the crystal. At the ideal defect-free (100) surface one O^{2-} ion is assumed to be missing in the direction of the surface normal and the clusters contain only five O^{2-} ions in a semi-infinite point-charge Madelung field. The O_h symmetry of the bulk of the transition-metal oxide is reduced to the tetragonal C_{4v} symmetry at the surface.[5] As can be directly inferred from the simple one-electron picture of Fig. 2.3, the remaining degeneracy of the d states is further lifted at the surface because the change in the crystal field, caused by the missing oxygen ion in the z-direction (surface normal), affects mainly the d orbitals with a z component, especially the $3d_{z^2}$ orbital. Whereas the energy required for the $3d_{xy} \rightarrow 3d_{x^2-y^2}$ excitation remains unchanged, the $3d_{xz} \rightarrow 3d_{z^2}$ and $3d_{yz} \rightarrow 3d_{z^2}$ excitations are energetically lowered, leading to a further splitting of the crystal-field multiplet terms [44]. A part of the crystal-field multiplet of the surface Ni^{2+} ions in NiO

---

[5] Cr_2O_3 has the corundum structure. For bulk 3d multiplet calculations, the cluster used consists of one Cr^{3+} ion, surrounded by a slightly distorted octahedron of six O^{2-} ions. If the (111) surface is considered, the cluster contains the surface Cr^{3+} ion and the three O^{2-} ions, representing its next neighbors [13, 128].

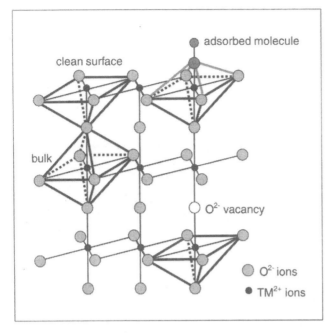

clean surface

adsorbed molecule

bulk

O^{2-} vacancy

O^{2-} ions

TM^{2+} ions

Fig. 2.8. Illustration of the differences in symmetry of the crystal field experienced by bulk and surface transition-metal ions. Bulk ions are octahedrally surrounded by six O^{2-} ions. At the clean (100) surface (*left*), one O^{2-} ion of the octahedron is assumed to be missing and the surface transition-metal ions are tetragonally surrounded by five O^{2-} ions only, leading to a C_{4v} symmetric crystal field. Bulk transition-metal ions located next to an O^{2-} vacancy are exposed to a very similar crystal field. Surface TM ions adjacent to molecules adsorbed at regular surface sites (*right*) experience a nearly O_h-symmetric crystal field, similar to that of bulk TM ions (see Sect. 2.3.3.2)

is shown in Fig. 2.9 according to the calculations of Freitag et al. For Co^{2+} ions in CoO, a variety of surface d states has also been calculated [73,181]. Calculated and measured surface d–d excitation energies of CoO and NiO are discussed in detail in Sects. 5.5.3, 5.3.2.2 and 5.5.4 (Table 5.4) together with our experimental results. At present, neither calculations nor measurements of the 3d multiplet of surface Mn^{2+} ions in MnO have been published to our knowledge.

As can be inferred from Fig. 2.8, the crystal field at bulk TM ions adjacent to a defect (O^{2-} vacancy) is also different from that of bulk ions in an undisturbed environment and can be expected to be very similar to that of the surface TM ions. In fact, the calculated multiplet d states differ only slightly in energy and it is hardly possible to distinguish between TM ions at the surface or at a defect [44,48]. The decision as to whether a measured d–d transition which cannot be attributed to a bulk excitation should be assigned to a surface or to a "defect" d–d transition must be made in close

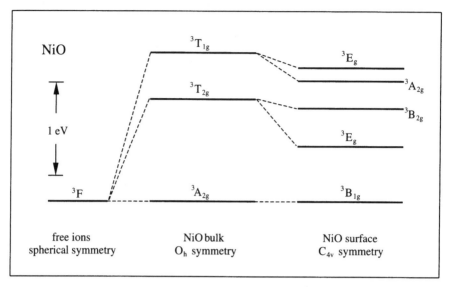

Fig. 2.9. Calculated crystal-field splitting of the 3d states of surface Ni^{2+} ions in NiO [44]. Only the splitting of the d states arising from the ^3F ground state of free ions is shown; calculations of the surface splitting of other terms of the d^8 configuration (Fig. 2.7) have not been published. The energetic positions of the bulk 3d states correspond to our experimental results (Sect. 5.5.4, Table 5.3)

relation to experimental results: transitions within the 3d multiplet of a bulk TM ion located next to an oxygen vacancy should not be influenced by the state of the surface, whereas all surface d–d transitions are very sensitive to surface contamination and adsorbates owing to adsorbate-induced changes in the surface crystal-field multiplet.

2.3.3.2 Spectroscopy of the Surface 3d Multiplet – an Efficient Tool for Studying Surface Properties and Adsorption. The knowledge of the surface's electronic structure and its interaction with adsorbates is essential for understanding catalysis. Thus, the growing interest in catalysis has recently forced the intensification of investigations of the electronic structure of transition-metal oxides surfaces. In particular the 3d states of the surface TM ions play an important role, since it has been discovered that the spectroscopy of these states – in combination with calculations – allows conclusions about the arrangement of surface ions, adsorption-induced changes, and adsorption sites to be drawn.

If a molecule is adsorbed[6] next to a transition-metal ion on a regular surface site, the adsorbate replaces the missing oxygen ion at the surface. The symmetry and strength of the crystal field become similar to that of the bulk

---

[6] The adsorption itself can be demonstrated by measuring the vibration frequencies of the adsorbed molecules by high-resolution electron energy-loss spectroscopy (Sect. 3.1).

at the adsorbate-covered surface (Fig. 2.8), and the excitation energies of the surface d–d transitions are shifted towards the bulk d–d excitation energies [44, 48]. Adsorption at surface defects on the other hand, has negligible influence on the surface 3d multiplet, and the surface d–d excitation energies remain unchanged. As already mentioned in the introduction (Chap. 1), the adsorption sites of several molecules have been investigated by use of this effect: NO is found to adsorb at regular sites on the NiO(100) surface, for example, whereas OH molecules, which cover the surface of several oxides after dissociative adsorption of water, are located at surface defects on the NiO(100) surface [17, 44, 46, 48]. On the NiO(111) surface, on the contrary, OH seems to be bound to regular surface sites [17].

The (111) surfaces of the transition-metal oxides are very reactive towards adsorption of small molecules [18] and are therefore of central interest concerning catalysis. The NiO(111) and CoO(111) surfaces, which can only be prepared as thin films on a substrate and not by cleavage, are found to be OH-terminated just after preparation, owing to adsorption and dissociation of H_2O molecules from the residual gas [18, 72]. The OH molecules stabilize the otherwise thermodynamically unstable polar (111) surface [19, 46, 72, 169]. Owing to the fact that the OH molecules occupy regular sites on the NiO(111) surface, the surface transition-metal ions are in an octahedral environment similar to that of the bulk (Fig. 2.8), and surface d–d excitations are only weakly indicated in electron energy-loss spectra. If the OH is removed by heating, the ions of the now unstable surface rearrange in order to reach a favorable energy and the surface reconstructs – as indicated by a change in the LEED pattern. In the energy-loss spectra, surface d–d excitations now become observable. From the comparison of these spectra with the very similar ones of clean NiO(100) surfaces and the results of cluster calculations for Ni^{2+} ions in various crystal fields, it is concluded that the reconstructed NiO(111) surface contains mainly Ni ions in a tetragonal C_{4v}-symmetric crystal field, typical of the (100) surface (Sect. 2.3.3.1, Fig. 2.8), but also a small amount of Ni ions in an environment with threefold symmetry [19, 48].

Major attempts have been made to determine the structure of the $Cr_2O_3(111)$ surface. Several possible positions of the surface Cr ions have to be considered here, each characterized by a slightly different crystal field provided by a different environment of oxygen ions. From a comparison of measured surface d–d excitation energies with calculated values obtained by cluster calculations, it seemed to be possible to determine the sites actually occupied by surface TM ions [13, 47]. But new results from the same group show that this might be impossible at the moment, because the calculated excitation energies for Cr ions at the different surface sites are too similar to allow a unique assignment [128]. High-resolution electron energy-loss measurements, which are briefly introduced in the next chapter (Sect. 3.1), and further improvement of the calculations may be helpful here.

In the field of catalysis, investigations of the adsorption of alkali metals are of interest, because alkalis can be used as promoters in catalytic reactions involving oxides. By monitoring the surface crystal-field changes, the adsorption of Na on $Cr_2O_3(111)$ has been studied [47, 48]: if a monolayer of Na is deposited, the d–d excitations of the surface Cr^{3+} ions, present for the clean Cr_2O_3 surface, vanish in the electron energy-loss spectra and two new excitations become visible. By comparing the excitation energies with results of cluster calculations, these authors attribute the results to d–d excitations of surface Cr^{2+} ions, indicating that the surface is covered with Cr^{2+} ions instead of Cr^{3+} in the presence of Na.

3. Electron Energy-Loss Spectroscopy – Inelastic Electron Scattering

3.1 Introduction

Electron energy-loss spectroscopy (EELS) is a widely used experimental technique for investigation of the excitation spectra of atoms, molecules, and solids. In this method, advantage is taken of the possibility of *excitation by electron impact*: incident electrons of fixed primary energy are inelastically scattered at the target, and the energy distribution of the scattered electrons is measured with respect to the incident electron energy. This energy distribution – the electron energy-loss spectrum – directly reflects the target excitations, because an excitation leads to the appearance of electrons in the scattered beam which have suffered a characteristic energy loss corresponding to the excitation energy. If other properties of the scattered electron beam (such as the angular distribution or polarization) are measured additionally (Sect. 3.2, 3.3), the kind of interaction between the incident electrons and the target that is responsible for the target excitation and the inelastic scattering process, can be inferred.

Electron energy-loss spectroscopy has its roots in the fundamental, early experiment of James Franck and Gustav Hertz [43]. This experiment, originally intended for the determination of ionization energies of vapors and gases, led to the discovery of excitations between discrete energy levels in mercury atoms and therefore provided the experimental proof of Bohr's postulate that forms the basis of modern quantum physics. With the development of efficient electron spectrometers in recent decades, electron energy-loss spectroscopy has advanced to an efficient tool for the investigation of excitations of atoms, molecules, and solids, requiring excitation energies from a few meV up to several hundred electron volts. Electron energy-loss spectroscopy and optical absorption spectroscopy are almost equivalent experimental methods to a certain degree, but in several respects EELS is superior to optical methods and is therefore an excellent complement to optical absorption spectroscopy. The special advantages of electron energy-loss spectroscopy are reviewed here briefly, with regard to the transition-metal oxides and related materials.

Apart from electronic transitions, collective vibronic excitations such as phonons and plasmons are observable with EELS. In addition, Auger transitions and vibronic excitations of adsorbed molecules can be investigated [89, 91, 167]. In the field of vibronic excitations in transition-metal oxides, longitudinal optical surface phonons due to interaction of the electrons with the ionic crystal lattice, the so-called Fuchs–Kliewer phonons, have been observed by means of electron energy-loss spectroscopy [9,24,44,45,208], [200, pp. 140ff], [19,72,73]. Bulk plasmons in transition-metal oxides [6,63,65,184], as well as vibration frequencies of several adsorbed molecules, have been measured (see below).

Generally, two kinds of electron energy-loss experiments have to be distinguished, differing mainly in the primary energy of the incident electrons and the kind of excitations investigated. In high-energy EELS, primary electrons with energies usually exceeding 100 keV hit a thin target and the energy distribution of the scattered electrons is measured in transmission. The energy resolution is of the order of several hundred meV or even worse. In the second type of electron energy-loss spectroscopy the impact energy is much lower (between several eV and a few keV) and the energy loss of the scattered electrons is measured in reflection. Therefore, this kind of spectroscopy is often called REELS (*reflection* electron energy-loss spectroscopy). The energy resolution usually ranges from less than 1 meV [90,92] (high-resolution electron energy-loss spectroscopy, HREELS) to ≈ 500 meV and is therefore better than in high-energy EELS experiments. It should be noted that the energy resolution of HREELS experiments is quite comparable to that achieved in optical absorption spectroscopy.

With high-energy EELS, dipole-allowed excitations from core levels are usually investigated, because, except for excitations accompanied by a high momentum transfer from the incident to the scattered electron, excitations due to impact of high-energy electrons follow the same selection rules as excitations by photons, i.e. the dipole selection rules. High-energy EELS therefore corresponds mainly to x-ray absorption spectroscopy. With increasing momentum transfer, the probability of other electric and magnetic multipole transitions increases, because the corresponding terms in the inelastic scattering cross section can no longer be neglected in the Born approximation. For high momentum transfer, electric monopole and quadrupole transitions have in fact been observed, even with high-energy electron energy-loss spectroscopy [7,68].

An important step towards the understanding of high-temperature superconductivity has been made with high-energy EELS: it is well accepted now that the existence of O 2p holes in the CuO_2 planes in $YBa_2Cu_3O_{7-\delta}$ and $La_{2-x}Sr_xCuO_4$ and their interaction with localized Cu 3d holes is essential for the appearance of high-temperature superconductivity in the perovskites [70]. In $YBa_2Cu_3O_{7-\delta}$ and $La_{2-x}Sr_xCuO_4$, the O 2p holes are provided by doping with additional oxygen or strontium atoms, which take electrons from the CuO_2 planes. These compounds therefore exhibit superconducting behavior only if the oxygen or strontium content exceeds a certain value or lies in a certain range ($\delta \leq 0.5$ [210] and $0.3 \geq x \geq 0.06$ [70]), otherwise they remain in the semiconducting phase, even at low temperatures. The central significance of the existence of O 2p holes for high-temperature superconductivity in $YBa_2Cu_3O_{7-\delta}$ and $La_{2-x}Sr_xCuO_4$ has been shown impressively with high-energy EELS by demonstrating the δ- and x-dependent correlation between the appearance of the O 1s–O 2p transitions ($\approx 528\,eV$ energy loss), indicating the presence of O 2p holes, and the occurrence of superconductivity [149].

Low-energy EELS is a very surface-sensitive method. Owing to the very small mean free path length of ≈ 0.1–$1\,nm$ for electrons of 30–100 eV energy [176], the incoming electrons are able to penetrate only the first few atomic layers of a solid. (The lattice constants of the transition-metal monoxides, for example, lie between ≈ 0.4 and $0.45\,nm$; see Table 2.1). In addition to bulk excitations, excitations into or between surface states can therefore be investigated by low-energy EELS. If high-resolution low-energy EELS is applied, the aforementioned Fuchs–Kliewer surface phonons, as well as vibronic excitations of adsorbed molecules, can be studied. By means of this method, the vibration frequencies of several molecules, such as NO, OH, OD, CO, CO_2, and O_2, adsorbed onto NiO, CoO, and Cr_2O_3 surfaces have been measured [9, 17, 19, 44, 45, 46, 47, 72, 73, 105, 106, 169, 175, 213]. Not only are such investigations useful for proving the fact of adsorption unambiguously, but also conclusions about the surface properties can often be drawn – in particular if HREELS is combined with other surface-sensitive experimental methods: OH molecules are adsorbed at defects of the NiO(100) surface (Sect. 2.3.3.2). Nearly no OH adsorption is found at the surface of freshly cleaved NiO(100) single crystals, indicating a very small number of defects. Epitaxially grown NiO(100) thin films, on the other hand, are OH-covered just after preparation, and a higher amount of defects must therefore be concluded here [17]. The OH termination of the polar NiO(111) and CoO(111) surfaces and the coincidence of OH removal and surface reconstruction (Sect. 2.3.3.2) has been monitored by a combination of HREELS and LEED [19, 46].

Further, the strength of the adsorbate–substrate interaction can be estimated from HREELS investigations of adsorbate vibrations. This can be done by temperature-dependent measurements: above the desorption temperature, which is a direct measure of the binding energy between the adsorbate and substrate, the vibronic excitations of adsorbed molecules are no longer visible in the spectra [9, 72, 105]. Also, measurements of the deviation between the vibration frequencies of adsorbed and gas-phase molecules can be used: the C–O stretching frequency of CO, adsorbed onto NiO(100) and (111) surfaces, for example, is found to deviate only slightly from that of gas-phase molecules, indicating a rather weak bond with the substrate. For CO on the equivalent CoO surfaces a slightly higher but also small deviation from the gas-phase frequency is measured, and a rather weak bond must also be concluded here. For NO on NiO and CoO the situation is quite different. Here, the stretching frequencies deviate considerably from the gas-phase values because a strong interaction, provided by a chemical bond between the adsorbate and substrate, weakens the N–O bond and therefore shifts the vibration frequency [175].

In contrast to optical methods and high-energy EELS, low-energy EELS is especially appropriate for the excitation and examination of dipole-forbidden transitions. The reason is the "breakdown" of dipole selection rules for low-energy electrons: the probability of electric monopole as well as higher multipole transitions and *excitations by electron exchange* (Sect. 3.2.3.2), is expected to increase with decreasing energy of the incident electrons [7,71,122]. Several dipole-forbidden core excitations – such as 4d–4f, 3d–4f, and 4p–5d excitations with $|\Delta J| > 1$ or $\Delta S \neq 0$, for example – have been observed in reflection low-energy electron energy-loss spectra of rare earths [122, 136, 146, 185, 186]. For a review, see the article by Netzer and Matthew [147]. In energy-loss spectra of transition-metal oxides, the quadrupole-allowed ($\Delta \ell = 2$) transition-metal 3s–3d transitions occur with considerable intensity [64, p. 134], [56, 184] (Sect. 5.6.1, Fig. 5.32, Table 5.5).

Low-energy EELS has been applied, in particular, to several investigations of the dipole-forbidden 3d–3d excitations (Sect. 2.3) in transition-metal oxides[1] and 4f–4f transitions in rare earths and their compounds [10, 29, 57, 62, 123, 133, 134]. Owing to the aforementioned surface sensitivity of low-energy EELS, not only the bulk d–d transitions (Sect. 2.3.1) but also the surface d–d transitions (Sect. 2.3.3) of the transition-metal oxides are observable (Sect. 5.3.2, Figs. 5.23–5.26; Sect. 5.5.4, Fig. 5.31, Table 5.4; and references in Sect. 2.3.3). With optical absorption spectroscopy [85, 94, 148, 164, 166, 168] only bulk d–d excitations are observable.

In Sect. 3.2 the different inelastic scattering mechanisms relevant to excitation by electron impact are briefly reviewed. Subsequent to these considerations, the more sophisticated version of low-energy EELS with polarized electrons that we have used, spin-polarized electron energy-loss spectroscopy (SPEELS), which is required for an unambiguous proof of electron-exchange excitations and for the examination of their behavior, is presented (Sect. 3.3).

3.2 Electron Scattering

3.2.1 Introduction

For excitations due to electron impact, three inelastic electron scattering mechanisms due to different interactions of the incident electron with the target, needing different theoretical treatments [8, 89, 99], are often distinguished:

---

[1] EELS investigations of the d–d transitions in transition-metal oxides are the main subject of this work; references are therefore found in the whole text, in particular in Sect. 2.3 and Chap. 5.

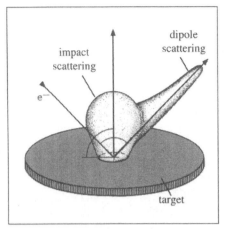

Fig. 3.1. Illustration of the angular distribution of dipole and impact scattering [115] (by permission of H. Ibach)

1. Dielectric dipole scattering (Sect. 3.2.2). This scattering is caused by the long-range interaction of the electron's electric field with the target charges and is excellently described in terms of classical dielectric dipole theory. The scattering cross section depends mainly on the dielectric function. A detailed microscopic knowledge of the interaction of the incident electron with the target is not needed. The angular distribution of the inelastically scattered electrons is very narrow, and the scattered electrons are confined to the so-called "dipolar lobe" in the forward direction in high-energy transmission EELS (Sect. 3.1) or the specular scattering geometry in low-energy reflection EELS if the energy loss and momentum transfer are small (Fig. 3.1, see also Sect. 3.2.2).
2. Impact scattering (Sect. 3.2.3). Impact scattering arises from short-range interactions between electron and target; the inelastically scattered electrons are usually found to be distributed over a wide angular range (Fig. 3.1). The term "impact scattering" is not defined precisely and covers nearly all scattering events not describable by classical dielectric dipole theory, such as exchange scattering, for example (Sect. 3.2.3.2). At the present, impact scattering is only poorly described by theory, owing to the necessity of more "microscopic" models, requiring a detailed knowledge of the electron–target interactions leading to the observed scattering process.
3. Resonant scattering via the formation and decay of a negative-ion compound state (Sect. 3.2.4). In this scattering mechanism, which can be regarded as a special form of impact scattering, an incident electron of suitable primary energy is captured into an unoccupied target state, forming a temporarily bound state. If the target is in an excited state after the decay of the compound state, the energy of the emitted electron deviates

from that of the incident electron by the excitation energy. The corresponding energy-loss peak in the EEL spectra is enhanced owing to the contribution of two excitation channels, the "normal" excitation, possible at any primary energy, and the excitation after decay of the compound state, possible at the resonance energy only. Resonant scattering has been found to be of central significance for the excitation of d–d transitions in NiO, CoO, and MnO for several primary energies (Sect. 5.2).

The classification of the scattering mechanisms by their different theoretical treatments may suggest a separate occurrence of these mechanisms in inelastic electron scattering. But one has to be aware that contributions of dipole- and impact-scattered electrons are usually superimposed in the measured electron energy-loss spectra. For special incident electron energies, contributions of resonantly scattered electrons are also found. As can be inferred from the following considerations, the measurement of additional parameters such as the angle or scattering-geometry dependence of the scattering process, changes in electron polarization during the scattering, or the primary-energy dependence must be used to identify the different scattering mechanisms and to determine their separate contributions to the inelastic scattering process.

3.2.2 Dielectric Dipole Scattering

Dipole scattering has been reviewed in detail by several authors (see, for example, the books and review articles by Ibach and Mills [89], Raether [167], Sturm [187], and Fink [41]). Only a few results relevant to the interpretation of our measurements (Chap. 5) will be summarized here: in dipole scattering, the target excitation is caused by the interaction of the electric field of the incident electron with the target. The electron itself is inelastically scattered owing to its interaction with the long-range dipolar electric fields arising from the charge fluctuations, in vibronic as well as electronic transitions in the target. But only those transitions with an electric dipole moment parallel to the electric field lines of the incident electron can contribute to the scattering process. Therefore, only transitions allowed by dipole selection rules can be excited [89, pp. 63ff]. Dipole scattering with small momentum transfer to the target occurs far above the target surface [89, p. 14] at distances where the wave-function overlap of the incident and target electrons is negligible. Therefore, transitions excited via the dipole-scattering mechanism are not accompanied by electron exchange [88].

For the description of dielectric dipole scattering, a detailed microscopic knowledge of the interaction between the incident electrons and target atoms is not needed, because the cross section for this type of scattering can be fully described in the framework of macroscopic dielectric theory. In this theory, the differential cross section in the first Born approximation, σ_{diff}, depends only on the imaginary part of the macroscopic dielectric function ε and the momentum $\hbar q$ transferred to the target in the scattering process

((3.3), Fig. 3.2). σ_{diff} is a measure of the number of electrons scattered with energy loss $\Delta E = \hbar\omega$ into the solid angle Ω and is given by [41]

$$\sigma_{\text{diff}} = \frac{\mathrm{d}^2\sigma}{\mathrm{d}\Omega\,\mathrm{d}\Delta E} = \frac{\hbar}{(\pi e a_0)^2}\frac{1}{q^2}\Im\left(\frac{-1}{\varepsilon(\boldsymbol{q},\omega)}\right). \tag{3.1}$$

Here a_0 is the Bohr radius; $\Im(-1/\varepsilon(\boldsymbol{q},\omega))$ is called the *macroscopic loss function*; $\varepsilon(\boldsymbol{q},\omega)$ is often almost independent of $\boldsymbol{q}$ and is usually replaced by the optical dielectric function $\varepsilon(\omega)$ in calculations. For a given energy loss ΔE, then,

$$\sigma_{\text{diff}} \propto \frac{1}{q^2}, \tag{3.2}$$

and q^2 can be calculated using momentum and energy conservation:

$$\hbar\boldsymbol{q} = \hbar\boldsymbol{k}_0 - \hbar\boldsymbol{k}_1, \tag{3.3}$$

$$\Delta E = \hbar\omega = E_0 - E_1 = \frac{\hbar^2}{2m}(k_0^2 - k_1^2), \tag{3.4}$$

where E_0 and $\hbar\boldsymbol{k}_0$ denote the energy and momentum of the incident electron, and E_1 and $\hbar\boldsymbol{k}_1$ the energy and momentum of the scattered electron.

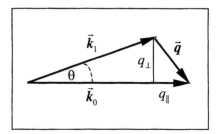

Fig. 3.2. Illustration of momentum transfer

In high-energy EELS (Sect. 3.1) the excitation energies of atoms and solids, and therefore the energy loss ΔE as well as the momentum transfer $|\hbar\boldsymbol{q}|$, are generally very small in comparison with the primary energy and momentum; $\Delta E \ll E_0$ and $|\hbar\boldsymbol{q}| \ll |\hbar\boldsymbol{k}_0|$. In this limit the scattering angle θ is also small and the scattering cross section (3.2) can easily be related to the scattering angle. For further calculations, it is useful to decompose the transferred wave vector $\boldsymbol{q}$ into two components $q_\parallel$ and $q_\perp$ parallel and perpendicular to the wave vector of the incident electron, $\boldsymbol{k}_0$. This is illustrated in Fig. 3.2 [167, p. 25], [41].

In the limit $\Delta E \ll E_0$ and $|\hbar\boldsymbol{q}| \ll |\hbar\boldsymbol{k}_0|$,

$$q_\perp = k_1 \sin\theta \simeq k_0 \sin\theta \simeq k_0\theta. \tag{3.5}$$

If, further, one takes into account that in this limit $q_\parallel = k_0 - k_1\cos\theta \simeq k_0 - k_1$ and $k_0 + k_1 \simeq 2k_0$, a simple transformation of (3.4) provides

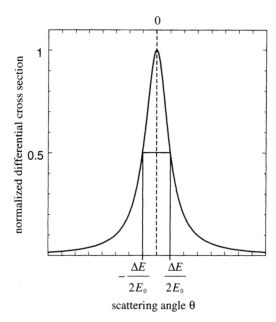

Fig. 3.3. Illustration of the dipolar lobe

$$q_\parallel \simeq k_0 - k_1 \simeq k_0 \frac{\Delta E}{2E_0}, \tag{3.6}$$

and the square of the transferred wave vector is then determined by

$$q^2 = q_\perp^2 + q_\parallel^2 \simeq k_0^2 \left[\theta^2 + \left(\frac{\Delta E}{2E_0} \right)^2 \right]. \tag{3.7}$$

Inserting this expression into the differential cross section for the inelastic scattering process (3.2) gives

$$\sigma_{\text{diff}} \propto \frac{1}{q^2} \simeq \frac{1}{k_0^2 \, [\theta^2 + (\Delta E/2E_0)^2]}. \tag{3.8}$$

This differential cross section has the shape of a Lorentz profile with a full width at half maximum (FWHM) of $2(\Delta E/2E_0)$. The intensity of the electrons scattered in an inelastic dipole scattering process is therefore expected to be confined to the so-called "dipolar lobe", which is strongly peaked in the forward direction. This is illustrated in Fig. 3.3, where the normalized differential cross section $\sigma_{\text{diff}}/\sigma_{\text{diff}}(\theta = 0)$ has been plotted versus the scattering angle θ.

The considerations leading to the differential cross section of (3.8) and Fig. 3.3 are strictly valid only if the energy loss and momentum transfer are small in comparison with the energy and momentum of the incident electron, as mentioned above. In transmission high-energy EELS, these conditions are always fulfilled, and experimentally the inelastically scattered electrons really are found to be highly concentrated in the small angular range around the incident beam direction predicted by (3.8). It might be expected that (3.8) could also be applied in low-energy electron energy-loss spectroscopy, where the primary energy is usually some orders of magnitude lower than in high-energy EELS (Sect. 3.1), provided the energy loss is not too high and the target transitions excited are dipole-allowed. In fact, the scattered electrons in reflection low-energy EELS are indeed often found to be mainly confined to a dipolar lobe, similar to that occurring in high-energy EELS [77, 78, 132, 209]. But this lobe is centered around the specularly scattered beam. In the specular scattering geometry of a reflection EELS experiment, the scattering angles are usually very large and can be close to 90° or even larger. From the dramatic decrease of the differential cross section for dipole scattering (Fig. 3.3) it is clear that a *single* large-angle dipole-scattering event cannot be responsible for the high intensities found at such large scattering angles. Here, instead, two scattering processes are assumed to be involved: the small-angle inelastic dipole scattering described above and a large-angle elastic scattering which may precede the inelastic process or occur after it [89, p. 69ff.], [162].

3.2.3 Impact and Exchange Scattering

3.2.3.1 Impact Scattering. Apart from the long-range dipole-scattering processes occurring in the vacuum far above the target surface (Sect. 3.2.2), electrons which reach the surface or penetrate into the solid can be scattered by short-range interactions during excitation of a target transition. These inelastic scattering mechanisms, which are not describable in terms of classical dipole theory, are usually summarized as what is called *impact scattering*. Impact scattering has barely been investigated theoretically up to now, because the theoretical treatment requires a detailed microscopic knowledge of the interaction between the electrons and the target. In contrast to dipole scattering, impact-scattered electrons are assumed to have a wider angular spread (Fig. 3.1)[2] and to be responsible for the often low but nonvanishing intensity observed at large scattering angles far away from the specular scattering geometry in low-energy EELS experiments [89, pp. 102ff.]. Differently from the dipole-scattered electrons in low-energy EELS, where two scattering events must be assumed to explain the intense dipolar lobe around the specular scattering direction (Sect. 3.2.2) – small-angle inelastic and large-angle

---

[2] It has to be noted that Fig. 3.1 gives only a rough, qualitative illustration of the angular distribution of impact-scattering processes. The maximum of the impact scattering cross section is found at different scattering angles for excitations in different materials and for different transitions. This maximum is usually not in the direction of the surface normal as indicated in Fig. 3.1. An isotropically distributed intensity arising from impact-scattering processes [78], as well as one increasing slightly towards the specular scattering geometry (scattering angle $\theta = 90°$), has been reported [88]. Slight maxima in the impact-scattering intensity, occurring in off-specular scattering geometries with grazing detection angles, are also observed [77].

elastic scattering –, large-angle impact scattering is often explainable by a single inelastic scattering event [129, 162].

In contrast to dipole scattering, impact scattering occurs very close to or even inside the target and can therefore be accompanied by electron exchange. It is this special kind of impact scattering, called *electron-exchange scattering*, which is found to be of central importance for the excitation of the d–d transitions in transition-metal oxides (Sects. 2.3.2, 5.2.3, 5.3). The present knowledge and expectations about electron-exchange scattering and excitation by electron exchange are summarized in the following paragraph, with respect to the d–d excitations.

3.2.3.2 Electron-Exchange Scattering. Electron exchange processes in the excitation of atoms and solids are well-known but poorly investigated phenomena. In particular, the angle and energy dependence of the exchange-scattering cross sections is not clear in detail, neither theoretically nor experimentally. The difficulties in the calculations of exchange-scattering cross sections arise from the necessity for detailed microscopic considerations of the electron–target interaction generally needed for the description of impact scattering, as already mentioned in Sect. 3.2.3.1. The reason for the lack of detailed measurements is found not least in the extreme experimental difficulties in identifying and investigating electron exchange: for a direct experimental check of exchange processes, it is necessary to distinguish between an ejected, exchanged, "true" target electron and an inelastically scattered incoming primary electron. Electrons "normally" are indistinguishable and become distinguishable only if their spin directions differ. Therefore, the only possibility of identifing and examining electron-exchange processes in nonferromagnetic materials directly is given by spin-polarized electron energy-loss spectroscopy with both a polarized primary electron beam and polarization analysis of the scattered electrons. This method, which we applied in our investigations of the d–d excitations in transition-metal oxides (Chap. 5), is described in Sect. 3.3. The experimental setup is introduced in Sect. 4.1.

Electron exchange seems to depend on the material and the specific kind of excitation. From experiments and calculations concerning free atoms and molecules, exchange is generally thought to be significant for primary energies which are not substantially larger than the excitation energy needed (typically up to ten times as large). The idea behind this belief is that exchange becomes more probable when the velocities of the free incident and bound target electrons involved in the collision are comparable [71]. Often, exchange processes are found to vanish when the primary energy exceeds the excitation energy only slightly: the well-known $6^1S_0 \rightarrow 6^3P_1$ transition in mercury, which was measured in the Franck–Hertz experiment [43], requires an excitation energy of 4.89 eV. This transition is found to be nearly exclusively excited by electron exchange for primary energies up to ≈ 7 eV. For higher primary energies, the cross section for excitation by exchange processes decreases strongly. Nearly no exchange excitations are detectable if the primary energy exceeds

$\approx 9.5\,\text{eV}$ [102, p. 130]. On the other hand, measurements on a ferromagnet (Co [93]) and earlier measurements of the d–d excitations of Cr_2O_3 [79, 80] and CoO [100] and of the f–f transitions of Gd and Eu [10, 29, 123, 133, 134] clearly show a high amount of exchange-scattering processes far above the excitation thresholds.[3] In our SPEELS investigations [50, 51, 52, 55, 56] of the d–d excitations in MnO, CoO, and NiO with primary energies up to 130 eV (Chap. 5), exchange was found to be significant even at these high energies, which exceed the d–d excitation energies (Figs. 2.5–2.7, Tables 5.1–5.3) by two orders of magnitude.

Stimulated by these results from our spin-polarized electron energy-loss measurements and earlier EELS results with unpolarized electrons [44, 65], Michiels et al. have calculated the energy dependence of the total impact/exchange-scattering cross sections of some d–d transitions in NiO recently for the first time [129], and the significance of electron exchange at high primary energies could also be shown. These calculations indeed contain detailed considerations of the microscopic electron–target interaction, as required for the description of impact scattering (Sect. 3.2.3.1). Here, space is separated into two regions: one where the scattered electron interacts fully with the Ni^{2+} ion of the target, and a second region outside, where the electron moves in an effective field provided by the target and the surroundings. In contrast to the theoretical description of electron scattering by free Ni ions, in calculations of electron scattering by Ni ions in NiO, the influence of the interaction between the Ni ion and the surrounding O^{2-} ions (Sects. 2.2, 2.3) must be introduced into the scattering potential. This is done in the calculations by taking the hybridization between oxygen ligands and Ni^{2+} ions (Sect. 2.2) into account by using a suitable crystal-field potential in addition to the Coulomb potential of the Ni^{2+} ion in NiO. For the latter, the Coulomb potential of free Ni^{2+} ions was scaled by a factor of 0.7. The total impact-scattering cross section was calculated for the $^3A_{2g} \rightarrow {}^3T_{2g}$ (^3F) $(\approx 1.1\,\text{eV})$

---

[3] The Gd and Eu measurements were performed by conventional EELS with unpolarized electrons: Gd and Eu have half-filled 4f shells ($4f^7$ configuration) with all spins parallel in the ground state. All f–f excitations are multiplicity-changing ones (similar to the d–d excitations of MnO, see Sect. 2.3). They are therefore expected to be excited by electron exchange exclusively (Sect. 3.3.3), and EELS measurements with unpolarized electrons are assumed to be sufficient for investigations of exchange scattering. That this is true and electron exchange is indeed the only excitation process for multiplicity-changing f–f transitions in rare earths was recently proved by our spin-polarized electron energy-loss measurements [57]. In the case of Co, a spin-polarized electron source but no spin analysis was used because only either spin-polarized primary electrons *or* polarization analysis of the scattered electrons is needed here, owing to the ferromagnetic order of the sample (Sect. 3.3.1). *Complete* SPEELS – with a polarized primary beam and polarization analysis of the scattered beam, as in the system that we use – has been applied in the case of Cr_2O_3 and CoO. Here, the slightly allowed multiplicity-conserving d–d transitions (Sect. 2.3.2) are expected to be also accessible to excitation via direct dipole scattering (Sect. 3.3.3) and a complete spin analysis is required to prove exchange unambiguously.

and the $^3A_{2g} \rightarrow {}^1T_{2g}$ (^1D) $(\approx 2.7\,eV)$ excitation in NiO (see Fig. 2.7 and Table 5.3) over a large incident energy range up to $\approx 50\,eV$. For both transitions, this cross section shows a similar, barely varying behavior in the energy range between $\approx 20\,eV$ and $54.4\,eV$. From the total cross section for the strongly dipole-forbidden multiplicity-changing triplet–singlet transition, containing exchange contributions only (Sect. 3.3.3), the important role of exchange excitations up to the high primary energy of $54.4\,eV$ can be directly inferred.[4]

Primary-energy-dependent exchange-scattering cross sections have also been calculated for Gd 4f–4f excitations recently. For primary energies of 100–300 eV, which are far above the excitation energies of a few eV, differential cross sections were found that were decreasing, but nonvanishing up to 300 eV [162]. The cross-section calculations of Ogasawara and Kotani [150] seem to confirm that the transition probability of the 4f–4f excitations in Gd^{3+} ions ($4f^7$ configuration) decreases rapidly with increasing incidend energy. Nevertheless, an EEL spectrum obtained from their calculated cross sections is found to be in good agreement with a measured spectrum [133] obtained from Gd metal with a relatively high incident energy of 145 eV.

Similarly to the case of the primary-energy dependence, the angular distribution of exchange scattering has barely been investigated up to now and is really not clear. The reason is that scattering-angle-dependent measurements of exchange-scattered electrons are extremely difficult, because not only are a polarized incident beam *and* energy and polarization analysis of the scattered electrons required, but the energy analyzer and polarization detector must be rotatable in addition. Therefore, in experiments the scattering angle is often fixed, and scattering-geometry-dependent measurements are made instead of true scattering-angle-dependent ones by simply rotating the target [78, 88, 209] (Sects. 4.1.1, 4.1.5). If multiplicity-changing transitions of free atoms or molecules or multiplicity-changing d–d or f–f excitations of solid transition-metal oxides and rare earths are investigated, electron exchange is often presumed to be the only possibility for the excitation of such transitions. Then the measurements are performed with unpolarized electrons, which requires the rotation of the energy analyzer only. The measured intensity of the scattered electrons or the differential cross section, is assumed in

---

[4] Besides, the cross section for this excitation has been calculated to be generally lower than that of the slightly allowed multiplicity-conserving triplet–triplet excitation (Sect. 2.3.2), which contains contributions from both, direct impact scattering (without exchange) and exchange scattering. This is in accordance with our experimental results: the energy-loss spectra of NiO and CoO are clearly dominated by the multiplicity-conserving transitions; the energy-loss peaks corresponding to the multiplicity-changing transitions are much weaker. But this is not only a consequence of the occurrence of direct impact scattering in the excitation of multiplicity-conserving transitions. In the specular scattering geometry this must be attributed mainly to contributions from excitations by the dipole-scattering mechanism (Sect. 5.3).

this case to correspond to the differential *exchange*-scattering cross section ([109], [16] and references therein, [64, 65, 123, 139]).

As previously illustrated in Fig. 3.1, exchange scattering, as a special type of impact scattering, is generally assumed to be much less dominated by forward scattering than dipole scattering is (Sect. 3.2.2, (3.8), Fig. 3.3) or is even assumed to be isotropically distributed. But as can be seen from the few existing experimental and theoretical results, this assumption is oversimplified, and large differences in the angular distribution of exchange-scattered electrons occur between various materials and different kinds of excitation. Exchange scattering is indeed nearly isotropically distributed for excitations in ferromagnetic Ni crystals [78] and for a multiplicity-changing singlet–triplet transition in CS_2 molecules [109]. Singlet–triplet excitations in He, on the other hand, lead to a pronounced maximum in the exchange-scattering differential cross section at small scattering angles. But, whereas the cross section corresponding to the He $1^1S \rightarrow 2^3S$ excitation strongly increases towards small scattering angles, the differential cross section for the scattering process leading to the $1^1S \rightarrow 2^3P$ excitation has a much broader angular distribution and reaches its maximum within a scattering-angle range of $10°$–$20°$ ([16] and references therein).

Concerning the angular dependence of exchange processes in dipole-forbidden transitions between localized states – such as the d–d excitations of transition-metal oxides or f–f excitations of rare earths and their compounds – only a few publications seem to exist. In the calculations of Porter et al. [162], the exchange-scattering processes leading to Gd 4f–4f and transition-metal oxide 3d–3d transitions are treated like inelastic electron-atom scattering. The differential exchange-scattering cross sections have been calculated for incident electrons of 100, 200, and 300 eV primary energy. They are found to increase symmetrically towards small scattering angles θ and to reach a maximum in the forward direction ($\theta = 0°$).

Detailed calculations by Michiels et al. [130] for a lower primary energy (20 eV), where the interaction between the TM ions and oxygen ligands has been introduced into the scattering potential (see above), predict a completely different behavior. In these calculations, performed for a fixed polar incident angle $\theta_i = 45°$ and fixed incident and detection azimuthal angles ($\Phi_i = 0°$ and $\Phi_d = 180°$, measured with respect to the (100) axis), the scattering angle θ is varied. The exchange-scattering cross sections corresponding to several d–d excitations in NiO are found to show a complicated scattering-angle dependence, which is strongly correlated with the symmetry of the 3d-multiplet final state (Table 2.3, Fig. 2.7): for excitations from the $^3A_{2g}$ ground state into different T_{1g} final states the cross sections are high for forward and backward scattering, but they exhibit distinct minima near scattering angles of $50°$ and $125°$, where the cross sections are very low. In contrast, the matrix elements and therefore the cross sections for the $^3A_{2g} \rightarrow {}^1A_{1g}$ excitations are found to vanish in the forward and backward directions ($\theta = 0°$ and $180°$).

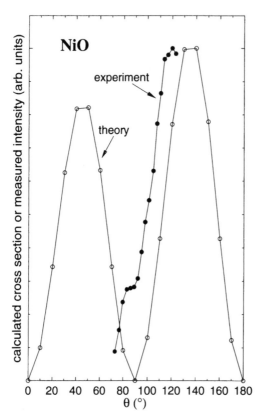

Fig. 3.4. NiO. Comparison of the angular dependence of the calculated exchange-scattering cross section of the $^3A_{2g} \rightarrow {}^1A_{1g}$ excitation (○) [130] and of the measured intensity of the 2.7 eV energy-loss peak (●) [139] (by permission of F. Müller and J. Inglesfield). The curves are normalized at the maxima. The intensity at 2.7 eV energy loss was obtained from the energy-loss spectrum by extrapolating and subtracting the "background" arising from the onset of transitions across the optical gap (see Sect. 5.1) linearly. In the calculation and measurements, the polar angles are measured with respect to the surface normal. The incident polar angle is fixed at $\theta_i = 45°$. The polar detection angle θ_d varies with the scattering angle θ, $\theta_d = (180° - \theta_i) - \theta$. The incident and detection azimuthal angles are fixed at $\Phi_i = 0°$, $\Phi_d = 180°$, both measured with respect to the (100) axis

The cross section is also zero in the specular scattering geometry ($\theta = 90°$, $\theta_i = \theta_d = 45°$) (Fig. 3.4). The calculations predict a constant ratio of spin-flip to nonflip scattering cross sections of 2, independent of the scattering geometry and incident energy, for the multiplicity-changing d–d excitations in NiO and CoO. The theoretical curve in Fig. 3.4 therefore represents the angular dependence of both spin-flip and nonflip scattering. Spin-flip and nonflip processes are described in detail in Sect. 3.3.

Recent true angularly resolved electron energy-loss measurements with unpolarized electrons (the experimental data in Fig. 3.4) – measured in the same scattering geometries as chosen in the calculations of Michiels et al. [130], seem to be in quite good agreement with the calculations, despite the slightly higher incident energy of 33 eV. The calculated cross section of the $^3A_{2g} \to {}^1A_{1g}$ excitation and the measured intensity of the 2.7 eV energy-loss peak, which is not only caused by the $^3A_{2g} \to {}^1A_{1g}$ (^1G) excitation but also contains contributions from the $^3A_{2g} \to {}^1T_{2g}$ (^1D) excitation (Table 5.3) owing to the energy resolution of 200–240 meV, exhibit a similar scattering-angle dependence. The shift of the experimental data towards smaller scattering angles in Fig. 3.4 is suggested to be due to the refraction of the electrons at the NiO surface: electron exchange processes causing bulk d–d excitations are supposed to occur within the sample (Sect. 3.2.3.1). On penetrating the surface, the incoming electrons are refracted towards the surface normal, and the outgoing electrons are refracted away from the normal. Therefore, the scattering angle outside the sample, which is always measured in the experiment, is smaller than the "true" scattering angle inside. The shoulder observed near the specular scattering geometry ($\theta = 90°$) in the experimental curve is attributed to the contribution of the $^3A_{2g} \to {}^1T_{2g}$ (^1D) excitation to the 2.7 eV energy-loss peak or to additional multiple scattering processes [139].

3.2.4 Resonant Scattering

3.2.4.1 Introduction. Resonance processes are well-known phenomena in photoemission spectroscopy as well as electron scattering, occuring when the energy of the incident particle is swept through the threshold for an inner excitation of the target. In both spectroscopic methods, the resonances occur because of the interference of two excitation channels leading to the same final state; the "normal" excitation process, possible at any energy of the incident particle, and excitation via formation and decay of a temporarily formed intermediate state, possible at certain resonant primary energies only. Resonant photoemission and resonant electron scattering are closely related processes, but they differ in some points, such as the nature of the intermediate and final states. Resonant photoemission has been described in detail elsewhere ([69, 196, 198] and references therein). Here, we shall concentrate on resonant electron-scattering processes, after some introductory remarks about those processes, occurring in transition-metal oxides and related materials (Sects. 3.2.4.2, 5.2.1).

Resonances in electron scattering are found in the elastic as well as in the inelastic total and differential scattering cross sections. Prominent examples are the resonances in the elastic-scattering cross sections of noble gases, which have been known and thoroughly investigated for more than thirty years [110]. Generally, the occurrence of resonant electron scattering is linked to the formation and subsequent decay of a temporarily formed negative-ion compound state.[5] Different kinds of the ion-formation process exist, which are traditionally used for the classification of the resulting resonance (see the review by Palmer and Rous [157], for example). Here, only the *core-excited* or *Feshbach* resonance will be described, owing to its relevance to the d–d excitations of the transition-metal oxides (Sect. 3.2.4.2): if the incident electron has a suitable energy, it can lose this energy in the form of a target excitation (often an excitation from a core level) and can itself be captured into an unoccupied target state. A compound state with one hole in an inner shell and two additional electrons in outer shells is formed. (An example of such a compound state is sketched in the left part of Fig. 3.5b, which is discussed in detail in Sect. 3.2.4.2.) This intermediate compound state can decay via an Auger process: an electron recombines with the hole created in the preceding excitation, and a further electron is emitted. After the decay of the compound state, the number of electrons is now again equal to that of the initial state; the target can now either be in the ground state or remain in an excited state.

In the case of an excited final state, the kinetic energy of the emitted electron is lower than the incident electron energy by an amount ΔE, equal to the energy of excitation into this final state. At the resonant primary energy, therefore, two excitation channels exist; excitation by creation and decay of a compound state (Fig. 3.5b), and direct excitation (Fig. 3.5a), which is possible at any primary energy of the incident electron. The interference of these two channels leads to a resonant enhancement of the inelastic electron scattering cross section and of the intensity of the corresponding energy-loss peak occurring at ΔE in the EEL spectrum. Resonances in the inelastic scattering cross section are observed not only if electronic transitions are excited, but also for excitation of vibrational and rotational transitions of free and adsorbed molecules ([77], [89, pp. 338ff]; [99, 112, 157] and references therein).

If the target atom or ion returns to the ground state after decay of the compound state, the emitted and incident electrons have the same kinetic energy. The formation and decay of the compound state then contributes to the elastic scattering, and its interference with direct elastic scattering processes leads to the resonant enhancement of the elastic scattering cross section.

---

[5] The term "negative-ion compound state" is used here in correspondence with the literature, where resonant electron scattering in the case of electron–atom or electron–molecule scattering is usually considered. It is clear that the compound state can be neutral or positively charged if the capture of the incident electron occurs at a positive ion, such as a TM^{2+} ion in a transition-metal oxide (Sect. 3.2.4.2).

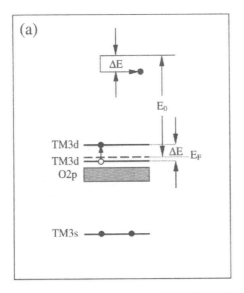

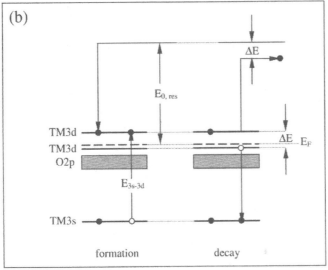

Fig. 3.5. (a) "Normal" d–d excitation (3.9a). The incident electron excites a d–d transition, requiring an excitation energy ΔE. An electron with a corresponding energy loss is emitted. This process is possible at any primary energy E_0. The d–d transition can be excited by an electron-exchange process or a direct scattering processes without exchange, such as direct impact scattering or dipole scattering (Sect. 3.2.2), if that is allowed owing to the relaxation of the dipole selection rules (Sect. 2.3.2). E_F denotes the Fermi level (Sect. 4.1.3). (b) Excitation by formation and decay of an intermediate compound state (3.9b). At the resonant primary energy $E_{0,\mathrm{res}}$ the incident electron can be captured into an empty 3d state during excitation of a 3s–3d transition; a compound state is formed. This state can decay by a Super-Coster–Kronig (Auger) process, where a 3d electron recombines with the 3s hole, transferring its energy to a further 3d electron, so that the target is left in an excited 3d state. The emitted electrons appear in the spectrum with an energy loss ΔE, corresponding to the d–d excitation energy

3.2.4.2 Inelastic Resonant Scattering in Transition-Metal Oxides and Rare Earths.
Resonances in inelastic electron scattering from materials containing localized states were observed for the first time in the 4f–4f transitions of rare earths when the primary-electron energy corresponded to the 4d–4f excitation energies [29, 133, 134], quickly followed by the observation of a similar resonance in the d–d transitions in NiO at a primary energy of 100–102 eV, which was attributed to the Ni 3s–3d threshold [65], [64, pp. 131ff.]. This resonance is described in detail below. The O 2p–Ni 3d excitation is also found to be resonantly enhanced at the 3s–3d threshold [65], [64, pp. 61ff.].

At the transition-metal 3s–3d threshold, the normal d–d excitation

$$3s^2 3d^n + e^- \rightarrow 3s^2 3d^{n*} + e^{-\prime}, \tag{3.9a}$$

which is possible at any primary energy, interferes with excitation via the formation and decay of a $3s^1 3d^{n+2}$ compound state, as described in Sect. 3.2.4.1:

$$3s^2 3d^n + e^- \rightarrow 3s^1 3d^{n+2} \rightarrow 3s^2 3d^{n*} + e^{-\prime}. \tag{3.9b}$$

Here n is the number of 3d electrons and $3d^{n*}$ denotes an excited 3d state. The notation $e^{-\prime}$ for the inelastically scattered electron is chosen to indicate the different energies of the incident and scattered electrons. These two transitions are illustrated in Fig. 3.5 (after [64, p. 133]).

The dipole-forbidden, but quadrupole-allowed 3s–3d excitations of the TM ions in transition-metal oxides are clearly visible in electron energy-loss spectra [64, p. 134], [56, 184] (Fig. 5.32, Table 5.5). The measured excitation energies $E_{3s\text{-}3d}$ are close to the binding energies of the 3s levels of the pure transition metals [76, p. 622]. In the case of NiO, the 3s–3d excitation energy (110.8 eV [184], 111 eV [64, p. 134]) exceeds the resonance energy of 100–102 eV considerably. This discrepancy has been explained [65], [64, p. 134] by the strong Coulomb attraction between the 3s hole and the additional 3d electrons which are located at the same Ni ion, owing to the localization of the 3d electrons in the transition-metal oxides (Sects. 2.2, 2.3). Nevertheless, the interpretation of the resonance was not uncontested. The situation has now been clarified by our measurements [55] on NiO, CoO, and MnO, which strongly affirm the assignment of the 102 eV resonance in NiO to the 3s–3d excitation: all the transition-metal oxides show this resonance, but with a shift of the resonant primary energy corresponding to the shift in the binding energy of the 3s level in the appropriate transition metal. Details are discussed in Sect. 5.2.1.

In our spin-polarized electron energy-loss investigations of the d–d excitations in NiO, CoO, and MnO a second, much stronger resonance is observed [51, 52, 54, 55, 56]. This resonance, which was found for the first time in that work, is described in detail in Sect. 5.2.1. It occurs when the energy of the incident electrons corresponds to the O 2p–O 3p excitation energy (36–38 eV).

Here, a compound state similar to that at the TM 3s–TM 3d threshold is assumed to be formed – the interference of the normal d–d excitation process with d–d excitations produced via decay of this compound state can lead to a resonant enhancement, as described above for the 3s–3d resonance ((3.9a,b) and Fig. 3.5). In particular, at this resonance at 36–38 eV primary-electron energy, the d–d excitations can be observed excellently: not only the dominant d–d excitations, which appear in the optical gap in the energy-loss spectra, but also those with excitation energies of several electron volts, which are strongly superposed by the dipole-allowed gap transitions, are clearly visible in the spin-resolved spectra in particular (Sect. 5.2.3.1, Fig. 5.10–5.12). By use of the O 2p–O 3p resonant primary energy, all sextet–quartet transitions (with one exception) predicted for the d^5 configuration in the O_h symmetric crystal field of MnO (Fig. 2.5) could be measured here (Sect. 5.5.2, Table 5.1), some of them for the first time to our knowledge. Several d–d transitions of higher excitation energy were also observed in CoO for the first time (Sect. 5.5.3, Table 5.2).

The formation and decay of a compound state leading to the resonant enhancement of a d–d excitation requires the capture of the incoming electron in the target. Therefore, resonant scattering contributes to the short-range impact and exchange scattering processes exclusively; the cross section for dipole scattering, where target excitation and the inelastic scattering process occur far above the target surface (Sect. 3.2.2), cannot be enhanced. From an intuitive point of view it is clear, in particular, that additional exchange channels are opened in the d–d excitation via the formation and decay of the compound state (Fig. 3.5b) because the captured electron forms a temporarily bound compound state together with the former target electrons and is no longer distinguishable from them. This intuitive assumption is strongly confirmed by our measurements, as discussed in Sect. 5.3.

3.3 Spin-Polarized Electron Energy-Loss Spectroscopy (SPEELS)

3.3.1 Introduction

The only possibility of direct experimental demonstration of electron-exchange processes in *nonferromagnetic* materials is provided by that kind of spin-polarized electron energy-loss spectroscopy that we have used: a polarized electron beam of known polarization P_0 is inelastically scattered at the target, and the intensity $I(\Delta E)$ as well as the polarization $P_s(\Delta E)$ of the scattered electrons is measured as a function of the energy loss ΔE. In contrast to the situation with *ferromagnetic* targets, where the use of either a polarized primary electron beam or polarization analysis of the scattered electrons is often quite sufficient for the experimental demonstration of exchange if the target magnetization can be reversed [78, 93], both the generation of a polarized electron beam and the use of an electron-spin detector are needed, making experimental investigations of exchange processes in nonferromagnetic solids much more difficult.[6] Details of our experimental setup are discussed in (Sect. 4.1).

If the spin–orbit interaction between the incoming electron and the target is negligible, as is the case for the 3d transition-metal oxides MnO, CoO, and NiO, which consist of lighter elements only (Sect. 5.3.2.1), the introduction of the electron spin as an additional parameter allows one to distinguish between exchange processes (Sect. 3.2.3.2) and direct excitations (without exchange) such as dipole scattering (Sect. 3.2.2) or direct impact scattering (Sect. 3.2.3.1) unambiguously: if the polarization $P_s(\Delta E)$ of the electrons scattered with energy loss ΔE deviates from the polarization of the primary electron beam P_0, a target excitation requiring an excitation energy equal to ΔE has been accompanied by an exchange of incoming and target electrons of opposite spin direction. Such exchange processes are usually called *spin-flip exchange* processes, although the spin of an electron is not flipped during the scattering process, but two electrons of different spin are exchanged. *Nonflip* exchange processes, where the exchange occurs between electrons of identical spin direction, are of course not detectable by SPEELS, because these electrons remain indistinguishable. Nonflip exchange processes therefore cannot be distinguished from direct scattering without exchange.

3.3.2 Spin-Flip and Nonflip Intensity

In addition to the spin-integrated energy-loss spectrum $I(\Delta E)$ provided by EELS with unpolarized electrons, spin-polarized electron energy-loss spectroscopy provides two further spectra, the spin-flip spectrum $F(\Delta E)$ and the spin-nonflip spectrum $N(\Delta E)$, which can be calculated from measurements of the spin-integrated intensity $I(\Delta E)$ and the polarizations of the incident and scattered electrons, P_0 and $P_s(\Delta E)$ [79]:

$$F(\Delta E) = \frac{1}{2}\left(1 - \frac{P_s(\Delta E)}{P_0}\right) I(\Delta E), \tag{3.10}$$

$$N(\Delta E) = \frac{1}{2}\left(1 + \frac{P_s(\Delta E)}{P_0}\right) I(\Delta E). \tag{3.11}$$

The spin-flip intensity gives the proportion of electrons in the total intensity $I(\Delta E)$ of the scattered electrons which was replaced by electrons of opposite spin direction during the scattering process. The nonflip intensity gives the proportion of electrons with unchanged spin direction.[7] These can

---

[6] In investigations of ferromagnetic materials, it is also the case that both polarized primary electrons and polarization analysis of the scattered electrons are needed if the spin-flip and nonflip intensities (Sect. 3.3.2) need to be determined separately [32, p. 27ff.], [209].

[7] In the case of ferromagnetic materials two spin-flip and two nonflip intensities are measured: one where the spin of the incident electron is parallel to the target

be scattered primary electrons or exchanged ones with identical spin direction. Therefore

$$I(\Delta E) = F(\Delta E) + N(\Delta E). \tag{3.12}$$

Direct excitations which are not accompanied by electron exchange, such as excitation via the dipole-scattering mechanism (Sect. 3.2.2) or direct impact scattering (Sect. 3.2.3), are expected to occur in the nonflip portion (3.11) of the energy-loss spectrum. The dipolar lobe (Figs. 3.1 and 3.3) is therefore expected to appear in the nonflip intensity exclusively. For dipole-allowed transitions, the dipole-scattering cross section usually exceeds the impact-scattering cross section by an order of magnitude or more in the vicinity of the dipolar lobe [89, pp. 113ff.], [8], as indicated in Fig. 3.1. The contribution of exchange scattering to the total scattering cross section can therefore be expected to be negligible near the specular scattering geometry in the case of dipole-allowed transitions; the spin-flip intensity (3.10) is very low in comparison with the spin-integrated intensity. But in scattering geometries other than specular, i.e. outside the dipolar lobe, spin-flip exchange processes can be of significance even for dipole-allowed transitions (see Sect. 5.3). The expectations for dipole-forbidden multiplicity-changing and multiplicity-conserving transitions are discussed in the following section.

3.3.3 Spin Flips and Nonflips in Dipole-Forbidden Transitions

Whereas in MnO, with its half-filled 3d shell, all d–d transitions are accompanied by a change in multiplicity, in NiO and CoO both multiplicity-changing and multiplicity-conserving d–d transitions are possible (Sect. 2.3.2). Both multiplicity-changing and multiplicity-conserving excitations are possible with and without change of the z component M_S of the total spin S of the transition-metal ion. This is illustrated in Fig. 3.6 for triplet–triplet and triplet–singlet excitations. The sublevels for different values of M_S and the possible transitions with $\Delta M_S = 0$ and $\Delta M_S = \pm 1$ are shown.

Exchange excitations of multiplicity-conserving as well as multiplicity-changing transitions are expected to be observed in both the spin-flip and the nonflip spectra. The reason is the relevance of the change in the z component of the total spin of the transition-metal ion (ΔM_S) to the appearance of an excitation in the spin-flip or the nonflip intensity. Exchange excitations with $\Delta M_S = 0$ are found in the nonflip intensity, and those with $\Delta M_S = \pm 1$ in the spin-flip intensity, independent of whether a change in multiplicity occurs. This is illustrated within the spin-vector model (Figs. 3.7 and 3.8) for selected triplet–triplet and triplet–singlet excitations from those shown in Fig. 3.6. In excitations with $\Delta M_S = \pm 1$, the incident and emitted electrons necessarily

---

magnetization, and one where they are antiparallel. These two spin-flip intensities as well as the nonflip intensities, are usually not equal [32, p. 63], [209]. This is a direct consequence of the different occupation of minority and majority states in a ferromagnet owing to the exchange splitting of the 3d bands.

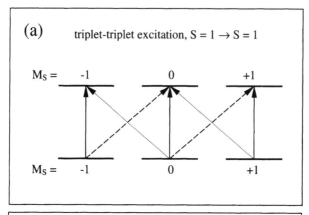

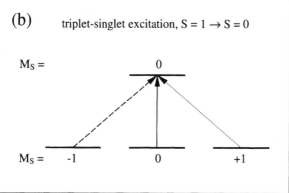

Fig. 3.6. Possible triplet–triplet (**a**) and triplet–singlet (**b**) excitations (NiO) with $\Delta M_S = 0$ (—), $\Delta M_S = +1$ (--), and $\Delta M_S = -1$ (···)

have opposite spin directions, as shown in Fig. 3.7a for a triplet–triplet and Fig. 3.8a for a triplet-singlet transition; such excitations appear in the spin-flip intensity exclusively. The only parity-forbidden transitions with $\Delta M_S = 0$ (Fig. 3.7b) appear in the nonflip intensity; the incident and emitted electrons have identical spin. These transitions are excitable by nonflip exchange and direct scattering (without exchange) additionally, owing to the identical spin configurations of the initial and final states. In addition to impact scattering, dipole scattering must be expected to contribute to the direct scattering owing to the relaxation of the parity selection rule (Sect. 2.3.2).

Not trivial at first sight is the case of the spin-forbidden transitions with $\Delta M_S = 0$ (the triplet–singlet transition in Fig. 3.8b): despite the change in multiplicity, the z components m_s of the spins of the incident and scattered electrons are identical and such transitions are not accompanied by a spin flip. They appear in the nonflip portion of the energy-loss spectrum only. Nevertheless, they require an electron exchange and cannot be excited by di-

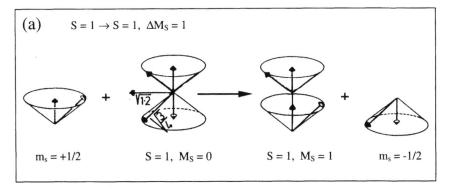

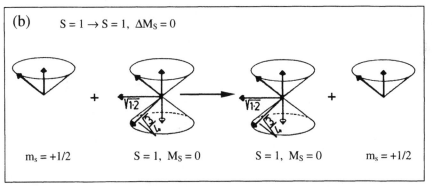

Fig. 3.7. Triplet–triplet transitions in the spin-vector model for incident spin-up electrons (after [102, p. 128]). (**a**) $\Delta M_S = 1$; (**b**) $\Delta M_S = 0$. The square roots located at several spins give the length of the spin vector in units of $\hbar$

rect scattering as the corresponding triplet–triplet transitions with $\Delta M_S = 0$ (Fig. 3.7b) can, because the final spin states are not identical. From a quantum mechanical point of view, triplet and singlet states with identical magnetic spin quantum number $M_S = 0$ have different spin wave functions (symmetric for triplet and antisymmetric for singlet states). In the semiclassical vector model the differences in the final states can easily be seen from a comparison of Figs. 3.7b and 3.8b: the addition of two single electron spin vectors with z components $m_s = 1/2$ and $m_s = -1/2$ to produce a total spin vector of zero length $(S = 0)$, as needed for the formation of a singlet state, requires spin-vector directions different from those needed for the total spin vector of the triplet state $(S = 1)$; although it has an identical z component $M_S = 0$, it has a finite length of $\sqrt{S(S+1)} = \sqrt{2}$ in units of $\hbar$.

Recent explicit calculations [130] of the differential spin-flip and non-flip cross sections of several d–d transitions in NiO and CoO show large differences for multiplicity-conserving and multiplicity-changing transitions (see Sect. 3.2.3.2): the cross sections for the spin-flip transitions have been calculated to be twice the cross sections for the nonflip transitions for the

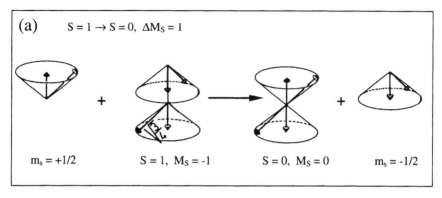

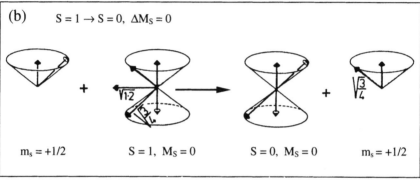

Fig. 3.8. Triplet–singlet transitions in the spin-vector model for incident spin-up electrons (after [102, p. 128]). (a) $\Delta M_S = 1$, (b) $\Delta M_S = 0$. The square roots, located at several spins, give the length of the spin vectors in units of $\hbar$

multiplicity-changing transitions, independent of the scattering geometry ($F/N = 2 \Leftrightarrow P_s/P_0 = -1/3$; see (3.10) and (3.11)). This corresponds to results obtained for the excitation of multiplicity-changing singlet–triplet transitions in free atoms with two electrons [102, p. 127ff.], where this factor of two is not only calculated but also measured for the $6^1S_0 \to 6^3P_1$ excitation of Hg, if the incident electron energy is close to the excitation energy of 4.89 eV (see Sect. 3.2.3.2). For the *multiplicity-conserving* transitions in these oxides, the situation is quite different, as has been calculated for the $^3A_{2g} \to ^3T_{1g}$ (^3F) excitation in NiO [130]. Here the nonflip cross section always exceeds that of the spin-flip scattering owing to the possibility of excitation by direct impact scattering (see above). The ratio of nonflip to spin-flip processes is not fixed; it varies strongly with the scattering angle.

4. Experiment

4.1 Experimental Setup

4.1.1 Introduction

The setup of our SPEELS experiment is shown in Fig. 4.1. It is described in this section briefly; a more detailed description of some parts of the apparatus is given in the following sections. The numbers in parentheses in the text correspond to the numbers in Fig. 4.1. The significant experimental parameters, such as energy resolution, primary polarization, and scattering angle, are summarized in Table 4.1.

Table 4.1. Significant experimental parameters. The energy resolution is the FWHM of the peak of the elastically scattered electrons

Primary polarization	$P_0 = 20\text{--}25\%$
Emission current	$I_e = 10\text{--}20\,\mu\text{A}$
Target current	$I_T = 100\text{--}500\,\text{nA}$
Primary energy	$E_0 = 20\text{--}130\,\text{eV}$
Energy resolution	$\Delta E_{1/2} \approx 230\,\text{meV}$
Scattering angle	$\theta = 90°$
Target temperature	$T = 400\text{--}500\,\text{K}$
Base pressure	$p < 2 \times 10^{-8}\,\text{Pa}$

Primary electrons longitudinally polarized in the z direction (Fig. 4.1) are generated by photoemission from a GaAs crystal (1) with circularly polarized light. A brief description of this process is given in Sect. 4.1.2. After an electrostatic 90° deflection (2), which takes the electrons out of the exciting laser beam (7), the electrons pass through a 180° spherical monochromator (4) (mean radius 50 mm; for the physical properties see [111]). Electrostatic deflection does not affect the spin direction. The polarization remains in the z direction and the initially longitudinally polarized electrons impinge transversely polarized onto the target.

The primary energy E_0 was varied between 20 and 130 eV in the measurements reported below. E_0 is always given with respect to the Fermi level of the oxide (Sect. 4.1.3).

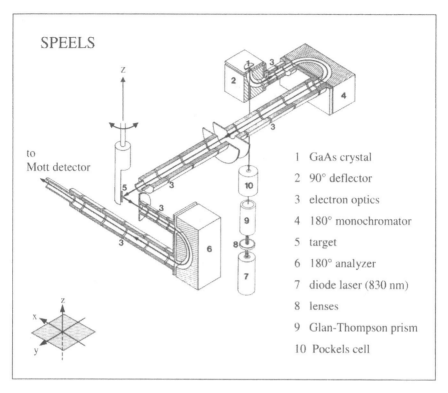

SPEELS

to
Mott detector

1 GaAs crystal

2 90° deflector

3 electron optics

4 180° monochromator

5 target

6 180° analyzer

7 diode laser (830 nm)

8 lenses

9 Glan-Thompson prism

10 Pockels cell

Fig. 4.1. Experimental setup ([32, p. 6])

The targets (5) were MnO(100), CoO(100), and NiO(100) single crystals. Measurements were carried out on freshly in-situ cleaved as well as on sputtered surfaces. The surface preparation, which differed slightly for the different oxides, is described in detail in Sect. 4.2. The target crystal is attached to a UHV manipulator, which allows exact positioning of the target for the measurements as well as for surface preparation and characterization.[1] The sample holder contains a heating element to keep the oxide target at 400–500 K during measurements and sputtering, to avoid charging (Sect. 4.1.4). At these temperatures, a sufficiently high conductivity could be obtained for incident electrons of more than 20 eV energy for NiO and of more than 28 eV and 30 eV for MnO and CoO, respectively.

Electrons scattered in the xy plane at a fixed angle of 90°, determined by the axes of the electron optics (3), are energy-analyzed by a spherical 180° spectrometer (6). This spectrometer is identical to the monochromator.

---

[1] The UHV chamber is equipped with a LEED/Auger system for surface characterization. However, the oxide samples were not examined by Auger spectroscopy, because impact of electrons of the high energies used in Auger spectroscopy was found to alter the surface stoichiometry of the oxides (Sect. 4.2).

Its acceptance angle is 1.5° [32, p. 8]. The total energy resolution of the apparatus, as given by the full width at half maximum of the elastically scattered electron peak, was ≈ 230 meV in the measurements reported below. Measurements in scattering geometries other than specular are possible by rotating the sample, which alters both the incidence and detection angles. Details of the scattering geometry are given in Sect. 4.1.5.

Spin analysis of the scattered and incident electrons is done by means of a conventional high-energy (100 keV) Mott detector (Sect. 4.1.6). For the measurement of the polarization of the incident electrons P_0, a repulsive electrostatic potential is applied to the target. Then the primary electrons are reflected into the Mott detector without interaction with the target atoms.

4.1.2 GaAs Source

For the generation of polarized electrons a GaAs source is used. Such sources, where polarized electrons are created by photoemission using circularly polarized light, have been described in detail elsewhere [158, 159, 160]. A brief description is as follows. GaAs is a direct-gap semiconductor with a minimal bandgap E_g at the Γ point ($E_g = 1.42$ eV at room temperature) [14]. Owing to spin–orbit interaction, the degeneracy of the valence band is partially lifted. The valence band is split into a fourfold $p_{3/2}$ and a twofold $p_{1/2}$ part, separated by the so-called "split-off" energy E_{so} ($E_{so} = 0.34$ eV at the Γ point) [14]. When the GaAs crystal (1 in Fig. 4.1) is illuminated with circularly polarized light (7–10) with $E_g < h\nu < E_g + E_{so}$, only transitions from the $p_{3/2}$ valence band to the $s_{1/2}$ conduction band near the Γ point are possible. Owing to the transition probabilities for these excitations with circularly polarized light, the excited electrons are up to $\pm 50\%$ polarized. The sign of the electron polarization depends on the sign of the photon spin [159, 160]. The highest polarization values are reached for $h\nu < E_g + 0.1$ eV [160]. Therefore, we use a diode laser (7) of wavelength 830 nm ($h\nu = 1.49$ eV).

The excited electrons thermalize into the bulk conduction-band minimum owing to phonon creation and inelastic scattering from valence-band holes [34, 37, 119]. Electrons in the conduction-band minimum which diffuse to the crystal surface before recombination with valence-band holes occurs can escape into the vacuum without being impeded by a surface barrier when the surface has a negative electron affinity (NEA). NEA conditions are achieved at the GaAs surface by lowering the vacuum level below the *bulk* conduction-band minimum (it always remains higher than the surface conduction-band minimum; see Fig. 4.2) by evaporating cesium and oxygen onto the surface. The NEA activation procedure is described in the review by Pierce et al. [160]. In contrast to the earlier work described by Pierce et al., we use a commercially available Cs dispenser, where Cs atoms are emitted owing to a chemical process when the dispenser is heated by resistive heating. To achieve good long-term stability of the GaAs photocathode, it is useful to maintain a weak cesium evaporation during the measurements.

Essential for the possibility of the formation of an NEA surface is a difference in the energy positions of the bulk and surface conduction-band minima, measured with respect to the Fermi level. NEA surfaces can therefore be achieved in p-doped GaAs crystals only.[2] In such crystals the acceptors provide a Schottky-barrier-like downward band bending of $E_b \approx E_g/3 \approx 0.5\,\mathrm{eV}$ at the surface [34, 172] (Fig. 4.2); the Fermi level is located close to the top of the bulk valence band [197, 205]. The work function Φ_s of the NEA GaAs source (Fig. 4.2) is then expected to lie in the energy range

$$\approx \frac{2}{3} E_g \leq \Phi_s < E_g$$
$$\Leftrightarrow \approx 0.95\,\mathrm{eV} \leq \Phi_s < 1.42\,\mathrm{eV}. \tag{4.1}$$

The emitted electrons are longitudinally polarized parallel or opposite to the direction of the momentum of the incident photons [160], but with a degree of polarization lower than 50%, owing to spin-relaxation processes occurring during and after thermalization of the electrons ([161] and references therein). The polarization of the emitted electrons is correlated with the "quality" of the GaAs crystal used. It can differ between different GaAs wafers, but crystals cut from the same wafer usually exhibit nearly identical polarization values. The activation procedure using cesium and oxygen is of minor influence on the polarization of the emitted electrons [49, p. 87ff], [35]. In all measurements reported below, the polarization of the emitted electrons was $P_0 = 20$–25%.

One great advantage of the GaAs source in comparison with other sources of polarized electrons, which often provide a higher degree of electron polarization [11, 20], is its high current. A current of several microamps is typical for an incident laser power of a few milliwatts. With the GaAs crystals used by us and an incident laser power of $\approx 10\,\mathrm{mW}$, measured at the GaAs surface, a photocurrent of 10–20 μA is obtained with a freshly cesiated and oxidized surface. This current decreases gradually to 1–2 μA over several days. After heating to 550–650°C for $\approx 10\,\mathrm{min}$ and a subsequent new NEA activation with cesium and oxygen, the initial current values are reached again, and this can be repeated several times.

A further advantage of GaAs sources is the possibility of reversing the polarization of the electron beam very easily by changing the circularity of the light, which is of great importance for the elimination of instrumental asymmetries in the polarization measurements (Sect. 4.1.6). In our experimental setup (Fig. 4.1), this is done by reversing the sign of the voltage applied to the Pockels cell (10).

Our GaAs source is equipped with a 180° spherical monochromator (4), which allows a reduction of the width of the energy distribution of the electrons impinging on the target. But owing to the low counting rates in

---

[2] Good GaAs photocathodes have an acceptor concentration N_A of the order of 10^{18}–10^{19} atoms/cm^3 [119, 160]. The crystals used here are Zn-doped with $N_A = 4 \times 10^{19}$ atoms/cm^3; the surface is (100)-oriented.

spin-resolved electron energy-loss spectroscopy of dipole-forbidden excitations (Sect. 4.2.1), a high current of incident electrons is desired. The energy distribution of electrons emitted from GaAs photocathodes is usually very narrow in comparison with that for thermionic cathodes and has a full width at half maximum of the order of ≈ 30–130 meV, if the energy of the incident light exceeds the bandgap only slightly.[3] During the measurements of the d–d excitations reported below, a fixed monochromator resolution of 200 meV was chosen. In this case the transmission of nearly all electrons emitted from the GaAs source can be expected.

4.1.3 Primary Energy

For the occurrence of the electron-exchange and resonant scattering investigated in the work described in this book (Chap. 5), the penetration of the incident electrons into the crystal must be assumed (Sects. 3.2.3, 3.2.4). Therefore, the primary energy E_0 *inside* the target crystal, after acceleration by the target work function, is relevant. This energy is determined mainly by the voltage V_a applied between the GaAs photocathode and the target. The energy relations between the GaAs and oxide crystals are illustrated in Fig. 4.2.

As can be inferred from Fig. 4.2, E_0 exceeds the energy difference eV_a between the Fermi levels of the source and target only slightly. Because of the lowering of the work function (4.1), the nearly complete thermalization of the excited electrons into the GaAs conduction-band minimum prior to emission, and the very narrow energy distribution of the emitted electrons (Sect. 4.1.2), the energy of all electrons emitted from the GaAs source is expected to lie between Φ_s and E_g with respect to the Fermi level of the source.

From Fig. 4.2 and (4.1), it is clear that E_0 deviates from the applied potential eV_a by

$$eV_a + \Phi_s \leq E_0 \leq eV_a + E_g$$
$$\Leftrightarrow \quad 0.95\,\text{eV} + eV_a \leq E_0 \leq 1.42\,\text{eV} + eV_a. \tag{4.2}$$

A more exact value of E_0 cannot be given, owing to the uncertainties in the work function of the source (4.1) and in the position of the maximum of the energy distribution of the emitted electrons. But, in the measurements below, where eV_a was varied between 20 and 130 eV, a deviation of the order of 1 eV

---

[3] The shape of the energy distribution and its FWHM depend on the energy of the incident light and the crystal temperature [34, 40, 95, 160]. In addition, a dependence on the work function Φ_s of the NEA surface is found. If the work function rises towards E_g ((4.1), Fig. 4.2), which can happen owing to nonoptimal preparation of the NEA surface or "natural aging" of the cathode by contamination with residual-gas atoms, the low-energy part of the energy distribution of electrons reaching the band-bending region of the GaAs crystal is cut off. The width of the energy distribution of the emitted electrons is then reduced [34, 39, 103].

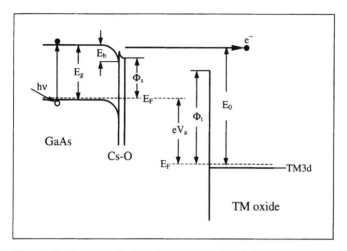

Fig. 4.2. Energy relations between an NEA GaAs source and a transition-metal oxide target. For the GaAs crystal, the top of the valence band and the bottom of the conduction band at the Γ point, the band-bending towards the surface, and the Cs–O layer (Sect. 4.1.2) are shown. For the oxide, only the highest occupied levels, the TM 3d states, located close to the Fermi level, are shown (for other states see Fig. 3.5)

between E_0 and eV_a (4.2) is negligible. Therefore, in all spectra presented in Chap. 5,

$$E_0 \simeq eV_a \qquad (4.3)$$

is taken as the primary energy of the incident electrons, while we must be aware of a possible energetic shift of the order of $\approx 1\,\mathrm{eV}$ inferred from (4.2).

Some remarks about the Fermi-level positions in Fig. 4.2. In the highly p-doped GaAs crystals used in GaAs photocathodes, the Fermi level lies very close (within several tens of meV) to the top of the valence band (Sect. 4.1.2). For the insulating transition-metal oxides, the position of the Fermi level is not well known. For NiO, it is known that this material always contains defects, and in fact it is thermodynamically stable only if it contains these defects (metal vacancies) [1,83], [84, p. 299]. At a metal vacancy, the charge of the missing metal ion is compensated by a hole on an adjacent oxygen ion. NiO therefore is a p-type material, with oxygen holes acting as acceptors. These acceptors determine the Fermi level, which is believed to lie within a range of 0.5 eV [83, 204],[84, p. 299] above the top of the "valence band" (containing band-like O 2p as well as localized Ni 3d states; see Sect. 2.2) and therefore to provide the conductivity at higher temperatures (Sect. 4.1.4). The existence of such defect acceptors and therefore a similar Fermi-level position must also be assumed for CoO and MnO, because both compounds show a high-temperature conductivity similar to that of NiO.

4.1.4 Target Temperature

For all experimental techniques involving charged particles used in the work described here (i.e. electron energy-loss spectroscopy, low-energy electron diffraction, and cleaning by sputtering with ions; see Sects. 4.1.1, 4.2), the samples under investigation must exhibit a sufficiently high conductivity to avoid charging. Wide-gap insulators with a very low conductivity such as the transition-metal monoxides are usually not accessible to these techniques. The key here is the slight deviation from stoichiometry, mentioned above in Sect. 4.1.3: at temperatures higher than room temperature, the acceptors arising from the metal vacancies provide a sufficiently high conductivity. For NiO crystals, heating to 400–450 K allows measurements with incident electrons of primary energies higher than 20 eV. For CoO and MnO slightly higher temperatures are needed (for CoO, ≈ 450 K, MnO, 450–500 K); measurements with electrons of primary energy E_0 higher than 30 eV and 28 eV, respectively, were possible here.

4.1.5 Scattering Geometry

The scattering geometry in the work described here is illustrated in Fig. 4.3.

As mentioned in Sect. 4.1.1, the scattering angle θ is determined by the axes of the electron optics (Fig. 4.1) and fixed at 90°. In the specular scattering geometry, therefore, the incidence angle θ_i and the detection angle θ_d are both 45° with respect to the surface normal (Fig. 4.3a). The scattering geometry can be varied by rotating the sample by an angle δ, which alters the incidence angle as well as the detection angle: $\theta_i = 45° + \delta$, $\theta_d = 45° - \delta$. In the scattering-geometry-dependent measurements reported below (Sect. 5.3), the spin-resolved energy-loss spectra or polarization curves are plotted against

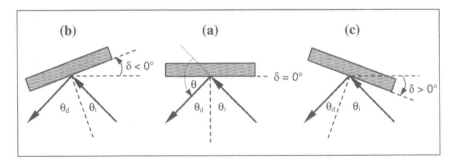

Fig. 4.3. Scattering geometry. The scattering angle is fixed at $\theta = 180° - (\theta_i + \theta_d) = 90°$; δ is the rotation angle of the target; $\theta_i = 45° + \delta$, $\theta_d = 45° - \delta$. (**a**) $\delta = 0°$, specular scattering geometry ($\theta_i = \theta_d = 45°$); (**b**) $\delta < 0°$, off-specular scattering geometry with rotation towards grazing detection angles ($\theta_i < 45°$, $\theta_d > 45°$); (**c**) $\delta > 0°$, off-specular scattering geometry with rotation towards grazing incidence angles ($\theta_i > 45°$, $\theta_d < 45°$)

the rotation angle δ. As illustrated in Fig. 4.3, $\delta = 0°$ in the specular scattering geometry. Negative δ values always correspond to a rotation towards more grazing detection and steeper incidence angles (Fig. 4.3b); positive δ values denote a rotation towards more grazing incidence and steeper detection angles (Fig. 4.3c).

4.1.6 Mott Detector

For the determination of the electron polarization, a conventional high-energy (100 keV) Mott detector is used in the apparatus described here. These commonly used detectors and their physical background are described in several publications; a brief survey is given in the book by Kessler [102, pp. 46ff. and 233ff.]. In summary, in these detectors advantage is taken of the spin dependence of electron–atom scattering due to spin–orbit interaction between the spin of the incident electron and its orbital angular momentum with respect to the scattering atom (Mott scattering). For materials with a high atomic number Z (a thin gold foil of 1000 Å thickness is used in the present system) the spin–orbit interaction leads to considerable differences in the differential scattering cross sections for transversely polarized electrons[4] with opposite spin directions. In the Mott detector this leads to an asymmetry in the counting rates N_r and N_l measured in two detectors placed symmetrically at scattering angles $\pm\theta$ in its scattering plane. The asymmetry for a totally polarized electron beam is called the Sherman function, S, and is a measure of the analyzing power of the spin detector. Apart from the Z dependence, the Sherman function depends strongly on the foil thickness, scattering angle, and electron energy. The polarization of an electron beam is given by

$$P = \frac{1}{S} \frac{N_r - N_l}{N_r + N_l}. \tag{4.4}$$

Our Mott detector follows the commonly used setup with five surface-barrier detectors, one in the forward direction for measurement of the spin-integrated spectra and two perpendicular pairs of detectors for the determination of the two possible transverse polarization components (in the z and y directions, Fig. 4.1). The detectors are placed at the angles usually chosen ($\theta = \pm 120°$), where the Sherman function of 100 keV electrons scattered at a gold target reaches its maximum [102, p. 64]. The Sherman function of the 1000 Å gold foil used in our Mott detector is $S = 0.21$ [32].

Instrumental asymmetries, which introduce errors into the polarization measurements, are always present in Mott detectors, owing to different efficiencies of the surface-barrier detectors and misalignments of the incident

---

[4] Owing to the dependence of the spin–orbit scattering potential on the scalar product of the spin and angular momenta, $\boldsymbol{\ell} \cdot \boldsymbol{s}$, only polarization components perpendicular to the scattering plane, i.e. transverse to the beam direction, can be measured.

beam. In our experimental setup, these asymmetries are eliminated by reversing the polarization of the incident electrons by changing the polarization of the exciting photons in the GaAs source (Sect. 4.1.2). If the helicity of the light is switched from σ^+ to σ^-, the polarization of the scattered electrons P_s changes its sign without a change in magnitude if instrumental asymmetries are negligible. On the other hand, if $|P_s(\sigma^+)| \neq |P_s(\sigma^-)|$, the instrumental asymmetry can be calculated and eliminated from the four counting rates obtained in the two detector pairs, for the two polarization directions of the incident electrons [102, pp. 233ff.], [36].

4.2 Targets and Target Preparation

All oxide target crystals were initially cleaved in air along the (100) plane from commercial, nominally single-crystal NiO, CoO, and MnO boules and then inserted into the UHV chamber. The further preparation of the surfaces prior to the measurements reported below differed slightly for the three oxides, owing to different cleavage behavior and differences in the sensitivity of the surface to chemisorption and damage by electron impact.

4.2.1 NiO

The NiO crystals were very easily cleaved in situ in the vacuum chamber. The surface of the in-situ cleaved NiO single crystals was of excellent quality; the low-energy electron diffraction (LEED) spots were very sharp. Below the Néel temperature of 523 K (Table 2.1), NiO showed the typical four half-order LEED spots (Sect. 5.4.1, Fig. 5.27), indicative of a multidomain antiferromagnetic order of the surface [74, 156]. The surfaces seemed to be very homogeneous; energy-loss spectra obtained from different areas of the surface were found to be identical.

The freshly cleaved NiO surfaces were very sensitive to damage by electron impact; the surface stoichiometry was altered even at low primary energies of 30–40 eV. After just 24 h of electron impact onto the same surface spot, the electron energy-loss spectra began to change slightly and the signal-to-background ratio of the energy-loss peaks asigned to the d–d excitations was reduced. The surface d–d excitations were especially strongly affected, because the excitation energies depend on the surrounding crystal field and therefore on the stoichiometry of the surface (Sect. 2.3.3). Reproducible measurements were possible for a few days after cleavage if the target position was changed with respect to the incident electron beam. Adsorption of residual-gas atoms or molecules seemed to be negligible up to several days after cleavage, because the spectra obtained from an area of the surface not exposed to electrons before were of the same excellent quality as those measured on a freshly cleaved surface. This insensitivity to contamination is attributed to the catalytic behavior of the surface.

The counting rates in spin-polarized energy-loss spectroscopy of dipole-forbidden transitions are usually very low. The measurements here were performed with rates down to less than 500 counts/s in the forward detector for the spin-integrated spectra (Sect. 4.1.6). The counting rates in the detectors for the backscattered electrons (N_r and N_l in (4.4)), which are needed for the polarization measurements, were of the order of less than 10 counts/s in this case, and the acquisition time needed to accumulate enough counts for significant statistics was comparatively long.[5] Owing to this very long data-acquisition time and the short "lifetime" of the freshly cleaved surface, it was not possible to measure on freshly cleaved NiO surfaces exclusively. Several measurements had to be performed on surfaces sputtered with 500 eV Ar ions for ≈ 30 min. The spectra obtained within two days after sputtering showed a sufficiently good correspondence with those obtained from freshly cleaved NiO crystals. Except for the surface d–d excitations (Sect. 2.3.3), which were clearly visible with freshly cleaved surfaces only, all d–d excitations were readily observable. However, they were slightly more visible in the spectra of cleaved surfaces, because the continuously distributed background due to sputter-induced defects was absent.[6]

4.2.2 CoO

CoO was cleaved in a similar simple manner to NiO. The CoO surfaces seemed to be less sensitive to surface damage due to electron impact, and the possible data-acquisition time for the same surface area was substantially longer; the counting rates for the d–d excitation energy-loss peaks were higher than in the case of NiO. Therefore, all CoO measurements shown in the following were obtained from freshly UHV-cleaved surfaces, but with several changes of the target position with respect to the incident electron beam, similar to those described for NiO (Sect. 4.2.1).

CoO seemed to be quite inert against adsorption. The energy-loss spectra of air-cleaved CoO were of very high quality, comparable to those obtained from the in-situ UHV-cleaved surfaces. All bulk d–d excitations were clearly observable. In air-cleaved NiO and MnO, on the other hand, these excitations were hardly visible.

4.2.3 MnO

The MnO samples were cleaved from a boule, also nominally single-crystal as in the case of NiO and CoO. But, in contrast to NiO and CoO, which showed

---

[5] For example, each data point of the primary-energy-dependent polarization curve for NiO (Fig. 5.13a, Sect. 5.2.3.2) took about 10–20 h data-acquisition time.

[6] A comparison of energy-loss spectra measured under identical scattering conditions on sputtered and freshly cleaved NiO crystals is presented in [50] and Fig. 5.26 (Sect. 5.3.2.2).

very homogeneous surfaces, the MnO cleavage surfaces were inhomogeneous and showed facets several square milimeters in area, indicating that the boule was not single-crystal, but actually consisted of several small crystallites. As reported earlier in the literature [75, p. 35], we also saw that cleavage of MnO was difficult. In contrast to the NiO and CoO single crystals used here, the MnO crystals could be cleaved in air only, and not in situ under UHV conditions. The cleaved MnO surface is known to be very active to chemisorption, which might be due to the formation of a large number of surface defects during cleavage [75, p. 220], and, in fact, the air-cleaved MnO surfaces showed very bad EEL spectra and had to be sputtered with 500 eV Ar ions and annealed at 450–500 K for several hours before spectra of good quality could be obtained.

Polished and annealed MnO surfaces are known to be of very good quality; sputtered and annealed surfaces have been found to be quite inert, especially in UHV [75, p. 220]. This leads to the conclusion that there is a very small number of surface defects for samples prepared by these techniques. In accordance with this assumption, the energy-loss spectra obtained from some of the surface facets of our sputtered and annealed samples showed an excellent quality, comparable to the spectra obtained from freshly UHV-cleaved NiO and CoO crystals. A continuously distributed background intensity in the optical gap, of the kind indicating surface damage in NiO (Sect. 4.2.1), was completely absent. But some hints of a slight deviation from stoichiometry were found for the MnO samples, because most energy-loss spectra obtained at various surface facets showed a weak excitation peak which was attributed to a d–d transition of Mn^{3+} ions in Mn_2O_3 (Sect. 5.5.2).

The sputtered and heated MnO surfaces in our experiments were also found to be very inert. Over several days, no change in the energy-loss spectra due to surface contamination was observed. In addition, damage due to electron impact, which limited the data-acquisition time in NiO strongly even at primary electron energies of less than 40 eV (Sect. 4.2.1), seemed to be nearly negligible for MnO in the primary energy range of up to 130 eV. After the prolonged initial sputtering of the MnO crystals described above, repeated sputtering for 0.5–1 h every three or four days provided reproducible energy-loss spectra of excellent quality.

5. SPEELS of Transition-Metal Oxides – Results and Discussion

5.1 Introductory Summary

In this chapter, the results of our spin-polarized electron energy-loss measurements of the dipole-forbidden d–d excitations CoO, and NiO are presented and discussed. Measurements of some transitions across the optical gap and of some dipole-allowed and dipole-forbidden excitations from upper core levels are also included. Most of the results and conclusions presented here have already been published [50,51,52,53,54,55,56]. References to these publications are often omitted in this chapter for simplicity.

The efficiency of electron energy-loss spectroscopy for the excitation and examination of dipole-forbidden d–d transitions may be demonstrated here, by way of introduction, by three excellent energy-loss spectra of the gap region of NiO, CoO, and MnO (Fig. 5.1). Details of the spectra, such as the assignment of the energy-loss structures to particular excitations or the gap width for example, are discussed below. The d–d excitations appear in the optical gap in the energy-loss range up to $\approx 6\,\mathrm{eV}$ and give rise to sharp energy-loss peaks, clearly demonstrating the localized, atomic-like nature of the participating initial and final states (Sect. 2.3). In contrast to optical absorption spectroscopy, where the intensities of d–d excitations and dipole-allowed transitions across the optical gap differ by several orders of magnitude (Sect. 2.3.2), in the energy-loss spectra here, the d–d excitation intensities are of the same order of magnitude as the dipole-allowed transitions, which give rise to a broad, more continuously distributed intensity occurring at energy losses above ≈ 4–$5\,\mathrm{eV}$. In particular, for MnO with its exclusively multiplicity-changing, parity- *and* spin-forbidden d–d transitions (Sect. 2.3.2), the d–d transition intensities exhibit astonishingly high values; the intensity of the dominant d–d excitation with $2.82\,\mathrm{eV}$ excitation energy and the intensities of the gap transitions are of nearly identical size here (Fig. 5.1c). As shown by our spin-resolved measurements, electron exchange is significant in the excitation of all the d–d transitions in NiO, CoO, and MnO with electrons of up to more than $100\,\mathrm{eV}$ primary energy.

The spectra of Fig. 5.1 were recorded with primary energies where the d–d excitations were found to be enhanced owing to the resonant scattering processes described in Sect. 3.2.4.2. It is the knowledge and use of the resonant primary energies in union with the special advantages of spin-polarized

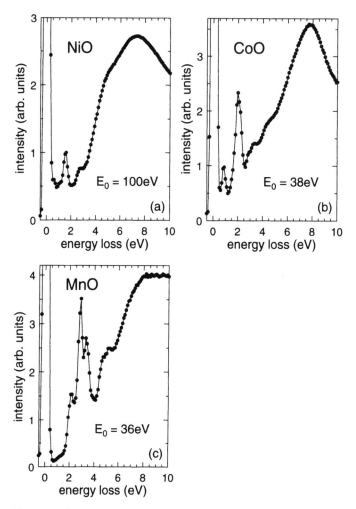

Fig. 5.1. Spin-integrated energy-loss spectra of (**a**) NiO, (**b**) CoO, and (**c**) MnO, measured at selected primary energies in the specular scattering geometry

electron energy-loss spectroscopy which is essential for the experimental determination of the excitation energies, their assignment to particular d–d transitions, and their comparison with calculated values: at off-resonant primary energies, all d–d excitation intensities are strongly reduced and only the dominant d–d excitations remain clearly visible; the weaker ones are hardly observable or not observable at all. The weaker excitations can be measured in resonance only (Sect. 5.2.1). Nevertheless, some of them remain barely visible even in resonance. But they can be clearly observed in the spin-resolved spectra: the d–d transitions with excitation energies above $\approx$3.5–4 eV in MnO and $\approx$2.5 eV in NiO and CoO are superimposed on the transitions across the optical gap, which increase rapidly with energy loss but are mainly confined to the nonflip intensity (Sect. 5.2.3.1). These d–d excitations and also transitions of low intensity which are superimposed on very intense ones are hardly visible in the spin-integrated spectra but give rise to distinct structures in the spin-flip spectra or in the polarization of the scattered electrons (Sect. 5.2.3.1) owing to the high contribution of spin-flip exchange processes to their excitation. By use of spin-resolved measurements in resonance, it is possible to measure and assign nearly all sextet–quartet d–d transitions of MnO. For CoO, some of the d–d excitations of higher excitation energies have been measured here for the first time. This is discussed in detail in Sect. 5.5.

Scattering-geometry-dependent spin-polarized electron energy-loss measurements (Sect. 5.3) are found to provide additional important information, not only about the contributions of different scattering mechanisms to the total scattering or excitation process (Chap. 3), but also concerning the assignment of energy-loss peaks to particular d–d transitions and the identification of surface d–d transitions. With these measurements, striking differences are found for the excitation of d–d transitions in and off resonance. *In resonance*, the excitation of multiplicity-changing as well as multiplicity-conserving d–d transitions is found to be completely determined by exchange processes in any scattering geometry; the intensity of the scattered electrons has a wide angular spread, but with a distinct maximum in the specular scattering geometry. *Off resonance*, the exchange scattering with a wide angular spread, is superposed on nonflip dipole-scattering processes, which are strongly confined to a small dipolar lobe around the specular scattering geometry (Sects. 3.2.2, 3.3.2), if slightly allowed multiplicity-conserving d–d transitions are excited (Sect. 2.3.2). Excitation by inelastic dipole scattering is nearly completely missing in the spin-forbidden, multiplicity-changing d–d excitations of MnO, as expected.

By means of scattering-geometry-dependent spin-resolved measurements, the dominant 2 eV d–d excitation of CoO (Fig. 5.1b), which has been a subject of controversy in the literature (Sects. 2.3.1, 5.5.3) and has often been assigned to a quartet–doublet excitation, has now definitely been identified as a slightly allowed quartet–quartet excitation ($^4T_{1g} \rightarrow {}^4A_{2g}$ (^4F)), because a considerable contribution of dipole-scattering processes to this excitation is found at off-resonant primary energies (Sect. 5.3.1.2).

The d–d excitations of surface Ni ions (Sect. 2.3.3), observable with freshly cleaved NiO crystals, and the bulk d–d excitations show completely different scattering-geometry dependences of spin-flip and nonflip intensities ((3.10) and (3.11)), providing a possibility to distinguish between them. Taking advantage of this possibility and the aforementioned high intensity of all d–d excitations at resonant primary energies, some surface excitations of NiO have been measured for the first time in the work described here (Sect. 5.3.2).

5.2 Primary-Energy Dependence and Resonances

5.2.1 Spin-Integrated Energy-Loss Spectra

Spin-integrated energy-loss spectra of the gap region of NiO are shown in Fig. 5.2 for selected primary energies between 25 and 115 eV. The intensities of all energy-loss peaks assigned to d–d excitations, which appear as distinct features in the optical gap, depend strongly on the primary energy and show a strong resonant enhancement around 38 eV and a weaker one at 100–102 eV primary energy. (A third resonance occurs around 30 eV primary energy. This resonance is discussed below.) Such resonances are also observed in CoO and MnO (Figs. 5.3 and 5.4): the d–d excitations of both CoO and MnO show a similar, strong resonant enhancement at the same primary energy of 36–38 eV and a second, weaker one at higher energy, at different energies for the different oxides (around 95 eV in CoO and 85–86 eV in MnO). The resonant enhancement of the energy-loss peaks arising from the $^4T_{1g} \rightarrow {}^4T_{2g}$ (^4F) and $^4T_{1g} \rightarrow {}^4A_{2g}$ (^4F) excitations of CoO (0.81 eV and 2 eV energy loss) and the

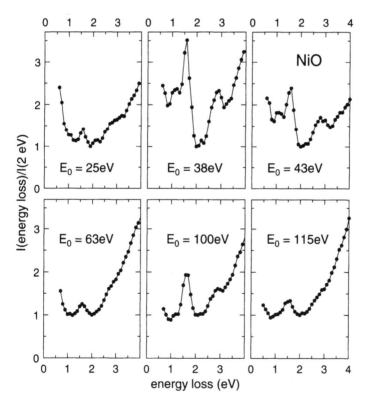

Fig. 5.2. NiO. Spin-integrated energy-loss spectra of the gap region, measured with different primary energies E_0 in the specular scattering geometry. The spectra are normalized to the intensity of the minimum at 2 eV energy loss

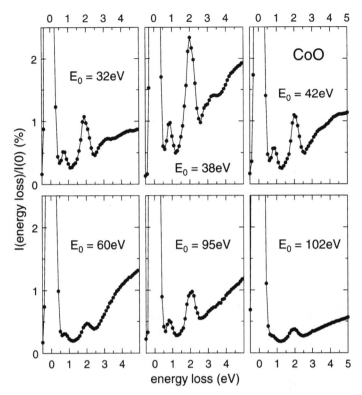

Fig. 5.3. CoO. Spin-integrated energy-loss spectra of the gap region, measured with different primary energies E_0 in the specular scattering geometry. The spectra are normalized to the intensity of the elastically scattered electrons

$^6A_{1g} \rightarrow {}^4A_{1g}$, ^4E_g (^{4}G) excitations of MnO (2.82 eV energy loss)[1] is shown in detail in Figs. 5.5 and 5.6a, where the intensities of the peaks, normalized to the intensity of the elastically scattered electrons ($\Delta E = 0$), are plotted versus the primary energy.

The resonances can be explained by resonant scattering processes of the kind described in Sect. 3.2.4. When the primary energy corresponds to an inner excitation threshold, the excitation of d–d transitions via formation and decay of an intermediate compound state interferes with the regular d–d excitations, which are possible at any primary energy, and the d–d excitations are resonantly enhanced. The existence of such processes requires the

---

[1] For the assignment of the energy-loss peaks to the d–d excitations several results of our primary-energy- and scattering-geometry-dependent SPEELS measurements (Sect. 5.2, 5.3) have been used. The assignments of the d–d excitation peaks are therefore discussed in detail below (Sect. 5.5) and are given here in anticipation of this discussion. The d–d excitation energies are average values, obtained from the evaluation of many of our energy-loss spectra. This is also discussed in Sect. 5.5.

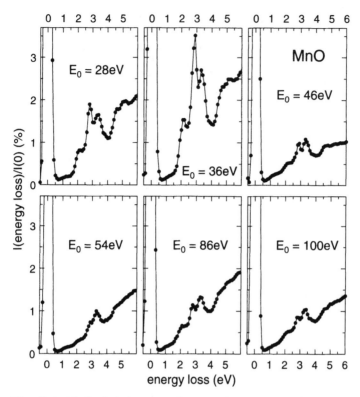

Fig. 5.4. MnO. Spin-integrated energy-loss spectra of the gap region, measured with different primary energies E_0 in the specular scattering geometry. The spectra are normalized to the intensity of the elastically scattered electrons

coincidence of the resonant primary energy with the excitation energy of another transition, such as an interband or core-level excitation. This coincidence is demonstrated in Fig. 5.6 for MnO. In addition to the primary-energy dependence of the intense $^6A_{1g} \rightarrow {}^4A_{1g}$, ^4E_g (^4G) excitations with 2.82 eV excitation energy (Fig. 5.6a), a part of an energy-loss spectrum, recorded with incident electrons of 130 eV primary energy is shown (Fig. 5.6b). The resonances (Fig. 5.6a) are correlated with the appearance of broad structures in the energy-loss spectrum (Fig. 5.6b), which we assign to the O 2p–O 3p and Mn 3s–Mn 3d excitation. These assignments are discussed in detail below and in Sect. 5.6.1 (Table 5.5). The slight shift of ≈ 2 eV between the maxima of the energy-loss structures (Fig. 5.6b) and the resonant primary energies (Fig. 5.6a) is a consequence of the process of formation of the compound state and the definition of the primary energy (Sect. 4.1.3): as can be inferred from Fig. 3.5b, for the resonance at the transition-metal 3s–3d threshold, the primary energy $E_{0,res}$ required for the formation of the com-

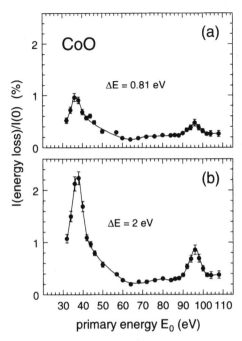

Fig. 5.5. CoO. Primary-energy dependence of the normalized intensity at (a) 0.81 eV energy loss ($^4T_{1g} \rightarrow {}^4T_{2g}$ (^4F) excitation) and (b) 2 eV energy loss ($^4T_{1g} \rightarrow {}^4A_{2g}$ (^4F) excitation), measured in the specular scattering geometry

pound state exceeds the 3s–3d excitation energy $E_{3s\text{-}3d}$ which is measured in the energy-loss spectrum (Fig. 5.6b).

In EELS of oxides, only the NiO resonance at high energy (≈ 100–$102\,\text{eV}$, Fig. 5.2) has been observed previously, and this was attributed to a resonant excitation process at the Ni 3s–Ni 3d threshold (Sect. 3.2.4.2, Figs. 3.5a,b and (3.9a,b)). As already noted in Sect. 3.2.4.2, this interpretation was not uncontested in the past, because, unlike our recent observations on MnO (Fig. 5.6), the resonance energy of 100–102 eV in NiO is shifted by several electron volts with respect to the 3s–3d excitation energy ($E_{3s\text{-}3d} \approx 111\,\text{eV}$, Sect. 3.2.4.2).

This shift has been explained [65], [64, p. 134] by the strong Coulomb interaction between the localized 3s hole and the two additional 3d electrons localized at the same Ni ion during the lifetime of the compound state (left-hand part of Fig. 3.5b). Strong interactions between the 3s core hole and the 3d electrons are also observed in photoemission spectra of NiO in the vicinity of the 3s threshold. There, the Coulomb interaction between the 3s hole created in the photoemission process and the 3d electrons lowers the energy levels of the Ni ion. An O 2p electron can be transferred into an initially empty 3d state, and the Ni 3s hole is screened owing to this "charge transfer" from the oxygen ligand. The energies required for photoemission from the Ni 3s

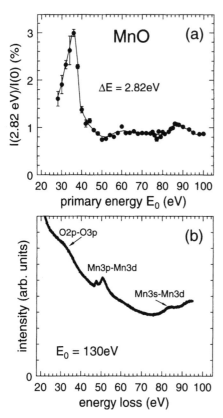

Fig. 5.6. MnO. (a) Primary-energy dependence of the normalized intensity at 2.82 eV energy loss ($^6A_{1g} \rightarrow {}^4A_{1g}$, ^4E_g (^4G) excitations), measured in the specular scattering geometry. (b) Energy-loss spectrum measured in the specular scattering geometry with 130 eV primary energy. The assignment of the energy-loss structures to the interband and core-level excitations is discussed in Sect. 5.6.1 (Table 5.5)

level with screening (O $2p^6$ Ni $3s^2$ $3d^8$ + $h\nu$ → O $2p^5$ Ni $3s^1$ $3d^9$ + e^-) and without this screening process (O $2p^6$ Ni $3s^2$ $3d^8$ + $h\nu$ → O $2p^6$ Ni $3s^1$ $3d^8$ + e^-) are different. A strong satellite, arising from this charge-transfer process[2], is observed in x-ray photoemission spectra at a binding energy 6.3 eV lower than the main line [200, pp. 80ff.]. However, although the observation of charge-transfer satellites in the photoemission spectra strongly supports the assumption of a strong 3s hole–3d electron interaction in NiO, the assignment of the 100 eV EELS resonance to the 3s–3d excitation remained doubtful,

---

[2] A detailed description of such charge-transfer processes and the corresponding satellites occurring in x-ray photoemission spectroscopy of several core levels in a variety of transition-metal compounds has been published by Veal and Paulikas [207].

owing to the large discrepancy between the resonance energy and the 3s–3d excitation energy.

Our measurements on CoO and MnO now strongly support this interpretation: if excitations between transition-metal states are exclusively responsible for the formation of the compound state involved in the resonance (Sect. 3.2.4.2), different resonant primary energies $E_{0,\text{res}}$ must be expected for the different oxides owing to the differences in the binding energies of the metal electrons. This is in fact observed (Figs. 5.3–5.6). In contrast to the low-energy resonance at 36–38 eV, which is discussed below, the high-energy resonance exhibits a distinct shift towards lower primary energy in CoO and MnO, in accordance with the lower binding energies of the 3s electrons in CoO and MnO.

The difference between the 3s–3d resonance energy ($E_{0,\text{res}}$) and the 3s–3d excitation energy ($E_{3s–3d}$) decreases in going from NiO to MnO. Whereas $E_{0,\text{res}} = 100$–102 eV and $E_{3s–3d} \approx 111$ eV in NiO (Sect. 3.2.4.2), $E_{0,\text{res}} = 95$ eV (Fig. 5.5) and $E_{3s–3d} = 101$ eV [184] in CoO. In MnO, the 3s–3d resonance primary energy corresponds exactly to the 3s–3d excitation energy (Fig. 5.6). Therefore, the perturbation of the TM^{2+} ion caused by the interaction of the d-electrons with the 3s hole localized at the same transition-metal ion seems to decrease with decreasing number of d electrons. A similar tendency is observed in 3s photoemission [200, p. 80ff.]. The above-mentioned splitting between the main line and the charge-transfer satellite, indicating the strong interaction between the 3s core hole and the 3d electrons in the photoemission spectra, is also found to decrease with decreasing number of d electrons. For MnO, with only five d electrons, the charge-transfer satellite is completely missing, indicating a negligible influence of the Coulomb attraction between the 3s hole and the 3d electrons on the position of the Mn^{2+} energy levels.

The strong low-energy resonance, occurring at nearly identical primary energies of 36–38 eV in the three oxides, is exactly correlated with a broad energy-loss structure (Fig. 5.6 for MnO, Fig. 5.7 for NiO and CoO), also appearing at nearly identical position (≈ 33–43 eV in NiO, ≈ 31–42 eV in CoO and ≈ 28–40 eV in MnO [2, 63, 98], [64, p. 72], [200, p. 61], [124]). The origin of this structure has been a subject of controversy in the literature, because several transitions may contribute to this structure, for energetic reasons: the O 2s–TM 4s/4p and O 2p/TM 3d–O 3s transitions have been suggested to be responsible for this loss structure [98], [200, p. 61], [124], as well as the O 2p/TM 3d–O 3p excitations [63], [64, p. 70]. The coincidence of the resonant primary energies (and the corresponding energy-loss structures) in the different oxides leads to the assumption that transitions between oxygen levels are responsible for their occurrence, owing to the nearly identical positions of the oxygen levels in the oxides [38, 211]. Therefore, we follow the argument of Gorschlüter and Merz [63], [64, p. 70] given for NiO and attribute the energy-loss structure around 38 eV mainly to transitions from the O 2p band, which

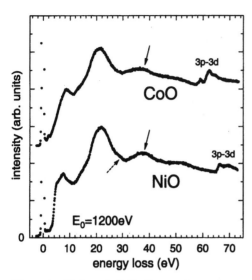

Fig. 5.7. NiO and CoO. Energy-loss spectra, measured in the specular scattering geometry ($\theta_i = \theta_d = 45°$) with 1200 eV primary energy ([64] p. 72, by permission of A. Gorschlüter). The CoO spectrum is arbitrarily shifted towards higher intensity. The O 2p–O 3p excitation (broad energy-loss structure around ≈ 36–38 eV), responsible for the occurrence of the strong low-energy resonance is denoted by *full arrows*. Even at this high primary energy, the intense d–d excitations (1.58 eV for NiO, 2 eV for CoO, compare Figs. 5.1–5.4) are weakly visible in the optical gap. The intense structures around ≈ 22 eV energy loss are attributed to the excitation of volume plasmons ([64, pp. 68ff.], [6,65]). The first transitions across the optical gap are discussed in Sect. 5.6. The *dashed arrow* denotes the O 2s–Ni 4s excitation (see text)

is located between ~ 2–8 eV below the Fermi level [38,58,59,211],[3] to an unoccupied band centered around 33 eV above the Fermi level. This band causes a broad maximum in bremsstrahlung isochromat spectra [64, pp. 32ff.]. It is attributed to O 3p states [64, pp. 35ff.] because x-ray absorption measurements [26] clearly show that transitions from the O 1s level into these final states are dipole-allowed, indicating their p-like character.

At the O 2p–O 3p threshold, the incident electron can be captured into an unoccupied 3d state during excitation of an O 2p electron into the O 3p band. An intermediate O $2p^5$ TM $3d^{n+1}$ O $3p^1$ state is formed. The Auger decay of this compound state into an O $2p^6$ TM $3d^{n*}$ final state, similar to that occurring in the TM 3s–TM 3d resonance (Sect. 3.2.4.2, Fig. 3.5b, (3.9b)),

---

[3] The photoemission peaks attributed to emission from O 2p and Ni 3d states overlap strongly. The Ni 3d peaks (mainly attributed to d^8L^- and d^7 final states) appear at binding energies only slightly different from the O 2p binding energy. The Ni 3d–O 3p transition may therefore require excitation energies similar to the O 2p–O 3p transition and may also contribute to the energy-loss structure around 38 eV.

explains the resonant behavior of the d–d excitations. In contrast to the 3s–
3d resonance, here the resonant primary energy is not shifted with respect to
the excitation energy. The reason is that the holes are not localized at the TM
ions in this case, and the strong interaction between the localized d electrons
and the hole, observed at the 3s–3d resonance in NiO and CoO, is missing.
Doubts about this interpretation of the 38 eV resonance arose from the fact
that interatomic Auger processes must be assumed, involving TM ions as
well as oxygen ions. A high probability of such processes seems doubtful
indeed, although the resonant behavior of the O 2p–Ni 3d transition at the
Ni 3s–Ni 3d threshold (see Sect. 3.2.4.2, (3.9a,b), but with a $3s^2$ O $2p^5$ $3d^9$ final
state) clearly shows the existence of interatomic Auger processes.[4] Besides, a
relatively high probability of interatomic Auger processes in MgO has been
shown by Matthew and Komninos [121]. It should be noted that all the
excitations which are suggested to contribute to the loss structure around
38 eV (see above) can cause a resonance only if interatomic Auger processes
are assumed.

In the transition-metal oxides of the 3d series, TM 3d and O 2p states are
hybridized (Sect. 2.2). This hybridization seems to be essential for the possibility of the interatomic Auger processes required for the resonant enhancement of the d–d excitations at the O 2p–O 3p threshold, as can be concluded
from recent primary-energy-dependent electron energy-loss measurements on
the f–f transitions of europium ions in europiumoxide [57]: similarly to the d
electrons of the transition-metal oxides of the 3d series (Sect. 2.2), the 4f electrons of rare-earth compounds remain localized at the rare-earth ions. But in
contrast to the 3d electrons of the transition-metal oxides, these 4f electrons
are effectively shielded by the filled 5s and 5p shells, which are higher in
energy. Therefore, they are hardly affected by the chemical environment and
retain nearly pure f character. In our electron energy-loss measurements with
primary energies up to 200 eV, only one weak resonance of the Eu f–f transitions was found. This resonance is attributed to the simultaneous excitation
of europium 4f–4f and 4d–4f transitions. Corresponding resonances have been
observed in a variety of pure rare earth metals [29, 133, 134] (Sect. 3.2.4.2).
Hints of further resonances in europiumoxide have not been found and a
resonance of the f–f transitions owing to simultaneous excitations involving
oxygen states must be excluded.

As can be inferred from Figs. 5.2–5.6, the intensity of the d–d excitations is
very high at the O 2p–O 3p resonance (around 36–38 eV) for the three oxides.
The TM 3s–TM 3d resonance is much less pronounced and barely visible in

---

[4] This resonant enhancement, previously shown by Gorschlüter [64, pp. 61ff.], was
also observed by us and leads to the high intensity at ≈ 7.3 eV energy loss (O 2p–
Ni 3d transition) in the spectra measured with the 100 eV resonant primary energy (Fig. 5.1a). At most other primary energies between 20 and 120 eV the intensity of the gap excitation, requiring ≈ 4.8–5 eV (shoulder in Fig. 5.1a), which
is assigned to the charge-transfer transition (Sect. 2.2), distinctly exceeds that
of the O 2p–Ni 3d excitation.

MnO (Figs. 5.4 and 5.6). The latter is surprising at first sight because the higher number of unoccupied 3d states leads to the expectation of a higher formation probability of the $3s^1 3d^{n+2}$ compound state (3.9b) in MnO owing to the higher number of possible channels for the 3s–3d excitation as well as for the capture of the incident electron into an unoccupied 3d state. In fact, the intensity of the TM 3s–TM 3d excitation is found to increase in the energy-loss spectra on going from NiO to MnO, according to this expectation [184]. But in the case of the resonances, not only must the 3s–3d excitation and the capture of the incoming electron, leading to the formation of the compound state, be considered, but also the subsequent Auger decay, responsible for the emission of electrons detected with energy losses corresponding to the d–d excitation energies, must be taken into account (Sect. 3.2.4.2). Here, the number of d electrons which can recombine with the 3s hole in the Auger process (Fig. 3.5b, (3.9b)) is higher in NiO and CoO than in MnO. Therefore the number of possible decay channels exceeds that in MnO, and this seems to be of relevance for the appearance of the stronger resonance in NiO and CoO.

In our measurements on NiO a third resonance is observed at $\approx 30\,\text{eV}$ primary energy. This resonance is explicitly demonstrated in Fig. 5.8, where further spin-integrated NiO spectra for a primary-energy range of 25–40 eV are shown. The resonant energy corresponds to a broad, weak structure between 27–31 eV in the energy-loss spectra [63, p. 72], [64] (Fig. 5.7), which is attributed to O 2s–Ni 4s excitations [64, pp. 69ff.] on the basis of the interpretation of bremsstrahlung isochromat spectra with the aid by calculations of Hugel and Belkhir [87] and the results of photoemission measurements by Wertheim and Hüfner [211]. A corresponding resonance is not observed in CoO and MnO, but this might be a consequence of the limited primary-energy range. In contrast to NiO, where primary-energy-dependent measurements were possible down to 20 eV, the measurements on CoO and MnO could not be extended to primary energies lower than 30 eV and 28 eV, respectively, owing to insufficient conductivity of the single crystals for incident electrons of lower energy (Sect. 4.1.4). Probably, O 2s–Co 4s and O 2s–Mn 4s excitations and the corresponding resonances occur at primary energies less than 30 eV; energy-loss peaks assigned to excitations involving transition-metal initial or final states are often expected to exhibit a shift of a few electron volts towards lower energy on going from NiO to MnO [124].

5.2.2 Other Correlations Between d–d Excitations and Transitions Across the Optical Gap?

The dipole-allowed TM 3p–TM 3d excitations appear as relatively sharp, distinct features in the energy-loss spectra of all transition-metal oxides [63, 98, 126], [64, p. 86ff.], [184] (Figs. 5.6 and 5.7). They usually consist of several components owing to the different Russell–Saunders terms of the $3p^5 3d^{n+1}$ final states (see Sect. 5.6.1 for MnO). But no resonant behavior of

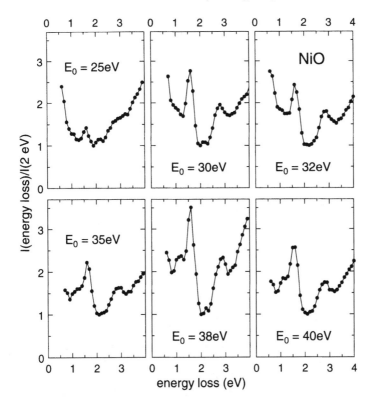

Fig. 5.8. NiO. Spin-integrated energy-loss spectra of the gap region, measured with primary energies between 25 and 40 eV in the specular scattering geometry. The spectra are normalized to the intensity at the minimum at 2 eV energy loss

the d–d excitations is found at the 3p–3d excitation threshold (Figs. 5.5 and 5.6). In photoemission, on the other hand, a resonant enhancement of several valence-band structures of the transition-metal oxides is observed at this threshold [60, 105, 113, 126, 178, 196]. But these resonances are rather weak [198]. In CuO, for example, the resonances observed at the 2p threshold are found to exceed those of the 3p threshold by two orders of magnitude [198]. Resonant photoemission experiments are therefore often performed at the 2p threshold [200, p. 27 and pp. 55ff.].

It is not clear why the 3p–3d resonance is completely missing in the energy-loss spectra. A reason why it could be expected to be weaker than the 3s–3d resonance in EELS might be the fact that 3d–3p transitions are dipole-allowed and the expected $3p^5\,3d^{n+2}$ compound state could decay by photon emission. An increased probability of radiative recombinations might lead to a decreased probability of Auger processes in the decay of the compound state. Then the number of emitted electrons would not be enhanced or would only be slightly enhanced, and a resonant behavior of the d–d excitation energy-loss peaks would not be observed. The 3d–3s transition required in the decay of the $3s^1 3d^{n+2}$ compound state, on the contrary ((3.9b) and Fig. 3.5b), is dipole-forbidden, and deexcitation of the compound state might preferentially occur by an Auger process, leading to the observed resonance at the 3s–3d threshold.

In MnO another, different correlation between d–d transitions and the Mn 3p–Mn 3d excitation seems to exist. The spin-integrated spectra (Fig. 5.4) show a striking primary-energy-dependent change in the peak ratio of the dominant d–d transition peaks at 2.82 eV and 3.31 eV excitation energy, which is not observed for d–d excitations of NiO or CoO (Figs. 5.2, 5.3, 5.8). This change seems in fact to be correlated with the Mn 3p–Mn 3d threshold (Fig. 5.9): when the Mn 3p–Mn 3d threshold is reached, the intensity of the 3.31 eV energy loss, which is assigned to the $^6A_{1g} \rightarrow {}^4T_{2g}$ (^4D) excitation (Sect. 5.5.2, Table 5.1) starts to increase and reaches a maximum at ≈ 10 eV above the Mn 3p–Mn 3d threshold. In contrast, the energy-loss peak assigned to the $^6A_{1g} \rightarrow {}^4A_{1g}$, ^4E_g (^4G) excitations (2.82 eV, Sect. 5.5.2, Table 5.1), which is the dominant energy-loss structure in the spectra recorded with 36 eV primary energy (Fig. 5.4), remains weak (Fig. 5.6a) and is hardly observable in the spectra obtained in this primary-energy range (Fig. 5.4, $E_0 = 54$ eV). The different behavior of these d–d excitations is clearly visible in Fig. 5.9b, where the ratio of their loss intensities is plotted:[5] at the O 2p–O 3p resonance at 28–40 eV primary energy, the ratio of the excitation peaks is low, owing to a weaker resonant behavior of the $^6A_{1g} \rightarrow {}^4T_{2g}$ (^4D) excitation, and increases drastically at the Mn 3p–Mn 3d threshold, owing to the prevalence of this excitation in this energy range (see Fig. 5.4, and also compare Fig. 5.6a with Fig. 5.9a). The peak ratio remains high up to ≈ 15 eV above threshold. Then it decreases strongly and reaches the value observed at the Mn 3p–Mn 3d threshold around 75 eV primary energy again. Towards higher energies it varies slowly. No distinct structures are observed at the Mn 3s–Mn 3d threshold, where the resonant behavior of both of the d–d excitations is found to be very weak (Figs. 5.6a and 5.9a). In earlier semi-angle-integrated electron energy-loss measurements, an increase of the peak ratio at the Mn 3p–Mn 3d threshold was also indicated but the ratio remained high and decreased slowly towards higher primary energy [98].

The origin of the differences in the d–d excitation cross sections, leading to the different strengths of the O 2p–O 3p resonance for the $^6A_{1g} \rightarrow {}^4A_{1g}$, ^4E_g (^4G) and $^6A_{1g} \rightarrow {}^4T_{2g}$ (^4D) excitations and the different behavior at the Mn 3p–Mn 3d threshold, is not yet clear. It might be owing to differences in the interactions between different excited final states of the 3d crystal-

---

[5] The intensity attributed to excitations across the optical gap increases strongly with increasing energy loss and gives a higher contribution to the energy-loss intensity at 3.31 eV than at 2.82 eV. Therefore, the energy-loss spectra have been fitted according to a procedure described in Sect. 5.5.2 (see Fig. 5.28) to obtain the intensities of the d–d excitations and gap transitions separately. In Fig. 5.9b, the intensity ratio of the fitted d–d excitation peaks is plotted.

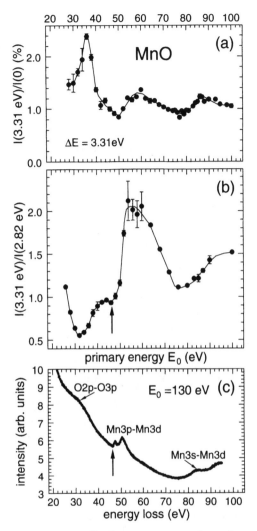

Fig. 5.9. MnO. Primary-energy dependence of (**a**) the normalized intensity at 3.31 eV energy loss ($^6A_{1g} \rightarrow {}^4T_{2g}$ (^{4}D) excitation) and (**b**) the intensity ratio of this peak to the 2.82 eV energy loss ($^6A_{1g} \rightarrow {}^4A_{1g}$, ^4E_g (^{4}G) excitations). (**c**) Energy-loss spectrum, measured in the specular scattering geometry with 130 eV primary energy (as in Fig. 5.6b). The *arrows* denote the onset of the Mn 3p–Mn 3d excitation

field multiplet and the intermediate compound state, as previously assumed by Jeng and Henrich [98] for the behavior at the Mn 3p–Mn 3d threshold. Calculations of such interactions do not exist, but it seems to be clear now that the crystal-field final state is of great influence on the excitation cross sections in EELS of transition-metal oxides. Recent calculations [130] of the spin-flip and nonflip differential cross sections of d–d transitions with final states of different symmetry in NiO ($^3A_{2g} \rightarrow {}^1T_{1g}$, $^3T_{1g}$ and $^3A_{2g} \rightarrow {}^1A_{1g}$) show strong symmetry-dependent differences (see Sect. 3.2.3.2). Interactions with transitions across the optical gap are not taken into account in these calculations. However, the different final-state symmetries of the $^6A_{1g} \rightarrow {}^4A_{1g}$, ^4E_g (^4G) and $^6A_{1g} \rightarrow {}^4T_{2g}$ (^4D) excitations in MnO might also be relevant to the different behavior of the corresponding energy-loss peaks (2.82 eV and 3.31 eV) at the Mn 3p–Mn 3d threshold and at the O 2p–O 3p resonance.

Also, an influence of the angular momentum is imaginable: the weaker energy-loss structure at 2.13 eV, which is assigned to a $^6A_{1g} \rightarrow {}^4T_{1g}$ (^4G) excitation (Sect. 5.5.2, Table 5.1), shows a primary-energy dependence similar to that of the 2.82 eV transition. The 2.13 eV excitation is hardly visible when the intensity of the 2.82 eV excitation is low but it increases strongly when the 2.82 eV peak increases in intensity (Fig. 5.4). Both of these energy-loss peaks are attributed to excitations into final states arising from d^5 terms of free ions (Table 2.2) with identical angular momentum (^4G), whereas the excitation with different behavior (3.31 eV) arises from a ^4D term.

From the measurements on MnO it is not possible to decide whether the symmetry or the angular momentum of the final states causes the differences, because all other d–d excitations with a corresponding symmetry or angular momentum are very weak or superposed on the strong excitations across the optical gap. But the measurements on CoO suggest that it is in fact the symmetry of the final state which is responsible for the different strengths of the resonances. Here, transitions into final states split from the same Russell–Saunders term of the free d^7 configuration (^4F) but with different symmetry in the crystal field ($^4T_{1g} \rightarrow {}^4T_{2g}$ and $^4T_{1g} \rightarrow {}^4A_{2g}$, with 0.81 and 2 eV excitation energy) also exhibit differences in the strengths of the O 2p–O 3p and Co 3s–Co 3d resonances. Interactions at the Co 3p–Co 3d threshold are not observed.

5.2.3 Spin-Resolved Measurements

5.2.3.1 Spin-Resolved Energy-Loss Spectra in Resonance. Spin-resolved energy-loss spectra of the whole gap region of the transition-metal oxides were measured at resonant primary energies in the specular scattering geometry. Only under these experimental conditions are the counting rates of the backscattered electrons in the Mott detector, needed for spin-resolved measurements (Sect. 4.1.6), high enough to obtain complete spectra of the gap region with sufficient statistical accuracy in a reasonably short time. Such spectra of MnO and CoO are presented in the upper parts of Figs. 5.10 and 5.11. The corresponding polarization P_s of the scattered electrons, normalized to the polarization P_0 of the incident electrons, is given in the lower parts of the figures. The data acquisition time needed for the spectra of Figs. 5.10 and 5.11 was ≈ 14–20 h and therefore short enough to neglect changes in surface conditions during the measurement owing to contamination or damage by electron impact (see Sect. 4.2). Off resonance and in off-specular scattering

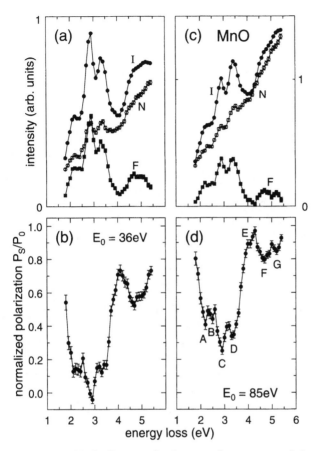

Fig. 5.10. MnO. Spin-resolved energy-loss spectra of the gap region, measured in the specular scattering geometry in resonance at 36 and 85 eV primary energy. (**a**), (**c**) spin-integrated intensity $I(\Delta E)$ (•), spin-flip intensity $F(\Delta E)$ (■), and nonflip intensity $N(\Delta E)$ (∘); (**b**), (**d**) normalized polarization P_s/P_0. The d–d excitations are labeled with capital letters (A–G) in (**d**)

geometries (Sect. 4.1.5) the counting rates were very low[6] and it was impossible to achieve complete spin-resolved spectra without repeated surface preparation during the measurement, i.e. sputtering of the MnO surface or change of the surface area exposed to the electron beam in the case of CoO (Sects. 4.2.2, 4.2.3).

A spin-resolved spectrum of NiO, measured at the 100 eV resonant primary energy, is shown in Fig. 5.12. In contrast to the conditions for CoO and

---

[6] The weak intensity of the energy-loss peaks assigned to d–d excitations at off-resonant primary energies can be inferred from Figs. 5.3–5.6. The intensity decrease in scattering geometries other than specular is discussed in detail in Sect. 5.3.

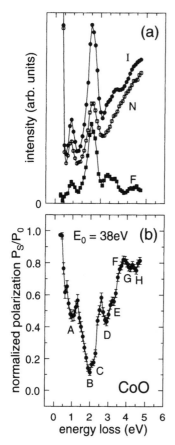

Fig. 5.11. CoO. Spin-resolved energy-loss spectrum of the gap region, measured in the specular scattering geometry in resonance at 38 eV primary energy. (**a**) spin-integrated intensity $I(\Delta E)$ (•), spin-flip intensity $F(\Delta E)$ (■), and nonflip intensity $N(\Delta E)$ (○); (**b**) normalized polarization P_s/P_0. The d–d excitations are labeled with capital letters (A–H) in (**b**)

MnO, it was impossible to achieve complete spin-resolved spectra of good accuracy with a freshly cleaved NiO surface or after a single sputtering process, owing to lower counting rates and an enhanced sensitivity of the freshly cleaved NiO surface to damage by electron impact (Sect. 4.2.1). Whereas the spin-integrated intensity I (Fig. 5.12a) was measured after one single sputtering process, the polarization curve was measured step by step over some weeks, with sputtering of the surface every second or third day. Each data point of the polarization curve of Fig. 5.12b is the average of several measurements, obtained on different days after different sputtering processes. The polarization values were found to coincide within the limits of the statistical error when the surface was well prepared.

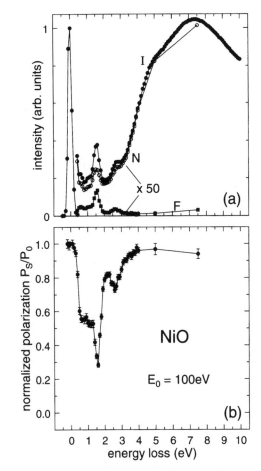

Fig. 5.12. NiO. Spin-resolved energy-loss spectrum of the gap region, measured in the specular scattering geometry in resonance at 100 eV primary energy. (a) spin-integrated intensity $I(\Delta E)$ (●), spinflip intensity $F(\Delta E)$ (■), and nonflip intensity $N(\Delta E)$ (○); the intensities are normalized to the intensity of the elastically scattered electrons. Above $\Delta E = 0.4$ eV, I, N, and F are scaled by a factor of 50. (b) normalized polarization P_s/P_0

As can be inferred from Figs. 5.10 and 5.11, all d–d excitation peaks of MnO and CoO are clearly reflected in the polarization curves (minima A–G in Fig. 5.10d and A–H in Fig. 5.11b). Owing to the large change in polarization in the scattering process, all d–d excitations occur in the spin-flip intensity (F in Figs. 5.10a,c and 5.11a), indicating a high contribution of spin-flip exchange processes in the quartet–doublet and quartet–quartet transitions in CoO and in all multiplicity-changing transitions in MnO. The aforementioned importance and advantage of resonant spin-resolved measurements for the identification of d–d excitations and the determination of their excitation energies (Sect. 5.1) can be seen in these spectra: the weaker d–d excitations, which are superposed on the strong dipole-allowed excitations across the optical gap, are clearly visible in the spin-flip intensity and in the polarization (F and G in Fig. 5.10d, F–H in Fig. 5.11b). Also, weak d–d excitations (B and E in Fig. 5.10d; C in Fig. 5.11b), which are invisible in the spin-integrated spectra owing to their superposition on very intense d–d excitations, can be resolved in the spin-flip intensity and the polarization curve. Details of the assignment of the energy-loss peaks to the d–d excitations and the determination of the d–d excitation energies are discussed in Sect. 5.5.2 and 5.5.3.

The elastically scattered electrons contribute to the nonflip intensity exclusively. This is clearly visible in the energy-loss spectrum of NiO (Fig. 5.12), where the spin-resolved measurements have been extended to excitations across the gap and to the elastically scattered electrons: the polarizations of the elastically scattered and primary electrons are equal ($P_s(0)/P_0 = 1$, Fig. 5.12b); N and I are indistinguishable near $\Delta E = 0$. In the CoO spectrum (Fig. 5.11), this behavior is also indicated; the flank of the peak of elastically scattered electrons appears in the nonflip intensity only; $P_s(\Delta E)/P_0 \to 1$ as $\Delta E \to 0$.[7] In our experimental setup the primary electron beam is transversely polarized, perpendicular to the scattering plane in any scattering geometry (Sects. 4.1.1, 4.1.5). For such scattering conditions, spin–orbit interaction is expected to alter the magnitude of the polarization vector (but not its direction) in the scattering process, except for particular scattering angles where the Sherman function might be zero [102, pp. 42ff.]. The absence of a change in the polarization of the elastically scattered electrons might therefore be indicative of a negligible spin–orbit interaction between the scattered electrons and the oxide targets. This is expected for the 3d transition-metal oxides, which consist of elements with low atomic number Z only, owing to the Z dependence of the spin–orbit interaction.

In the strongly increasing intensity attributed to the dipole-allowed transitions across the optical gap, nonflip excitations also prevail; most scattered electrons are found in the nonflip intensity, and the spin-flip intensity is small (Fig. 5.12, for NiO). In the CoO and MnO spectra, where the energy-loss range is limited to the gap region (Figs. 5.10 and 5.11), the prevalence of nonflip processes in the excitation of the gap transitions is indicated by the strong increase of the nonflip intensity and the normalized polarization P_S/P_0 towards higher energy loss. The high nonflip intensity is indicative of a large amount of direct scattering processes in the excitation of the gap transitions, i.e. direct impact scattering (Sect. 3.2.3) and, especially, dipole scattering, which is not accompanied by electron exchange at all (Sect. 3.2.2) and is

---

[7] The exclusive contribution of the elastically scattered electrons to the nonflip intensity was found for all oxides investigated here, independent of primary energy and scattering geometry. This can be seen, for example, by comparison of the NiO spectrum of Fig. 5.12 ($E_0 = 100\,\mathrm{eV}$), measured in the specular scattering geometry, with a spectrum published earlier, measured 20° off specular with 20 eV primary energy [50].

usually found to be the dominant scattering mechanism in the excitation of dipole-allowed excitations in the specular scattering geometry (Sect. 3.3.2).

5.2.3.2 Primary-Energy Dependent, Spin-Resolved Measurements of the Dominant d–d Excitations.

In Fig. 5.13, the primary-energy dependence of the normalized polarization in the three most intense energy-loss peaks of NiO, CoO, and MnO is shown for the specular scattering geometry. These are the 1.58 eV energy loss in NiO, which is attributed to the $^3A_{2g}$ → $^3T_{1g}$, ^1E_g (^3F, ^1D) excitations (Sect. 5.5.4, Table 5.3), the 2.0 eV energy loss in CoO ($^4T_{2g}$ → $^4A_{2g}$ (^4F) excitation, Sect. 5.5.3, Table 5.2) and the $^6A_{1g}$ → $^4A_{1g}$, ^4E_g (^4G) transitions with 2.82 eV excitation energy in MnO (Sect. 5.5.2, Table 5.1).

Such primary-energy-dependent spin-resolved measurements could be performed for the dominant energy-loss peaks only: as can be inferred from Figs. 5.2–5.4, weak d–d excitations are not observable at off-resonant primary energies. But even for the excitations of the highest intensity, the data acquisition time needed to accumulate enough counts for significant statistics in the polarization measurements was very long (Sect. 4.2.1). In the NiO curve (Fig. 5.13a), each data point took about 10–20 h measuring time. Thus, such measurements were possible on sputtered NiO surfaces only (Sect. 4.2.1). For the CoO crystals the counting rates were higher, and the CoO curve (Fig. 5.13b) could be obtained from an in-situ-cleaved crystal in ≈ 2–3 weeks. The surface area exposed to the electron beam was changed several times during the measurements, owing to surface damage by electron impact (Sect. 4.2.2). For MnO, where all measurements were performed on a sputtered surface owing to the poorer cleavage behavior (Sect. 4.2.3), the counting rates were comparable to those of cleaved CoO.

The d–d excitations of NiO and CoO investigated (Figs. 5.13a,b) exhibit a corresponding primary-energy dependence of the polarization: over wide primary-energy ranges, i.e. off resonance (Sect. 5.2.1), the polarization of the specularly scattered electrons deviates only slightly from that of the incident electrons; $P_s/P_0 \approx 0.8$. However, in resonance (30 eV, 38 eV, and 102 eV for NiO; 38 eV and 95 eV for CoO), a strong change in polarization is observed, indicating a high contribution of spin-flip exchange processes (3.10) even in the specular scattering geometry. In contrast, in MnO the polarization of the scattered electrons is low over the whole primary-energy range and deviates strongly from that of the incident electrons ($P_s/P_0 \ll 1$). The polarization difference in and off resonance is less marked than in NiO and CoO. The primary-energy dependence of the normalized polarization and the differences between MnO and the two other oxides (Fig. 5.13) can be understood only by means of spin-resolved scattering-geometry-dependent measurements. The interpretation of Fig. 5.13 is therefore discussed in the next section, together with these measurements.

The correspondence of the resonant primary energies of the O 2p–O 3p resonance in the three oxides, indicated by the decrease in polarization around

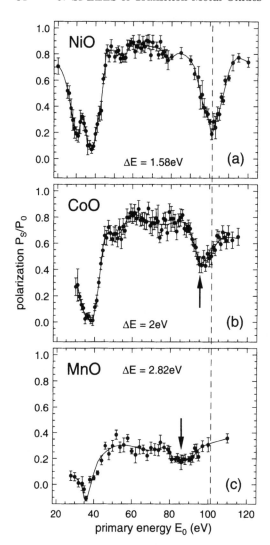

Fig. 5.13. Primary-energy dependence of the normalized polarization in the dominant d–d excitation peaks, measured in the specular scattering geometry: (a) NiO, (b) CoO, (c) MnO. The *dashed line* indicates the 3s–3d resonant primary energy in NiO. The *arrows* denote the TM 3s–3d resonances in CoO and MnO. The shift related to the reduction of the 3s binding energy is clearly visible

36–38 eV, and the shift of the resonant energies of the 3s–3d resonance, in accordance with the decrease in the TM 3s binding energies from Ni to Mn, is clearly visible in Fig. 5.13.

5.3 Scattering-Geometry Dependence

5.3.1 Bulk d–d Excitations

5.3.1.1 Introductory Summary. Scattering-geometry-dependent spin-in-tegrated energy-loss spectra of the gap region have been measured on NiO, CoO, and MnO for selected primary energies in and off resonance. The scattering-geometry dependence of the dominant, intense d–d excitation peaks, with excitation energies of 1.58 eV in NiO, 0.81 eV and 2 eV in CoO, and 2.82 eV in MnO (Figs. 5.1–5.4), has been investigated with spin reso-lution additionally. These energy-loss peaks appear near the middle of the optical gap and are therefore hardly superposed on excitations across the optical gap or on the flank of the peak of elastically scattered electrons. The combination of both spin- and scattering-geometry-dependent measure-ments can be used to distinguish between excitation by dipole scattering (Sect. 3.2.2) and electron-exchange scattering (Sect. 3.2.3.2): in resonance, both multiplicity-changing and multiplicity-conserving d–d transitions are found to be exchange-dominated in any scattering geometry. Off resonance, the electron-exchange scattering, which is still present, is superposed on strong nonflip dipole scattering, which is confined to a small dipolar-lobe-like feature around the specular scattering geometry (Sect. 3.2.2), if slightly allowed multiplicity-conserving d–d transitions of NiO and CoO are excited in the scattering process. Dipole scattering is found to be nearly completely missing in the strongly dipole-forbidden sextet–quartet excitations of MnO.

5.3.1.2 Multiplicity-Conserving Transitions in CoO and NiO. The scattering-geometry dependence of the bulk d–d excitations is excellently illustrated by three-dimensional plots (Fig. 5.14 for CoO) where the spin-integrated intensity is plotted against the energy loss and the angle of rotation δ. As previously stated in Sect. 4.1.5, the incidence and detection angles are given by $\theta_i = 45° + \delta$ and $\theta_d = 45° - \delta$. The specular scattering geometry corresponds to $\delta = 0°$. A selection of the spectra contributing to Fig. 5.14 is given explicitly in Fig. 5.30 (Sect. 5.5.3). A 3D plot for NiO (Fig. 5.24) is discussed in Sect. 5.3.2.2 together with the surface d–d excitations; 3D plots for MnO are shown in Fig. 5.20 (Sect. 5.3.1.3).

As can be inferred from Fig. 5.14 for CoO, the scattering-geometry depen-dence of the d–d excitation energy-loss peaks is completely different in and off resonance. The differences are clearly visible for the intense d–d transitions with 2 eV and 0.81 eV excitation energy (arrows in Fig. 5.14), which are the only d–d excitations clearly visible in energy-loss spectra of CoO measured at off-resonant primary energies (Fig. 5.3). In resonance (38 eV and 95 eV primary energy, Fig. 5.14a,c), the spin-integrated intensities of the d–d exci-tations are very high in the specular scattering geometry and decrease slowly when the sample is rotated. Off resonance (60 eV primary energy, Fig. 5.14b), the weaker spin-integrated intensity also reaches a maximum in the specular

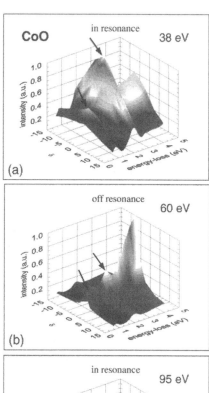

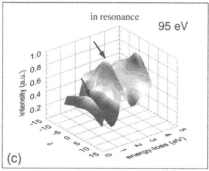

Fig. 5.14 a–c. 3D plots for CoO. The spin-integrated intensity is plotted versus the energy loss and the rotation angle δ of the sample. $\theta_i = 45° + \delta$, $\theta_d = 45° - \delta$. The curves for different primary energies are scaled in such a way that the different scattering-geometry dependences in and off resonance are clearly visible

scattering geometry, but decreases dramatically towards off-specular scattering geometries. The origin of this different behavior can be understood from the spin-resolved scattering-geometry-dependent measurements at 2 eV and 0.81 eV energy loss (Figs. 5.15–5.17): not only the spin-integrated intensity (I), but also the spin-flip intensity (F), as well as the nonflip intensity (N) and the normalized polarization P_s/P_0, are found to be completely different in and off resonance.

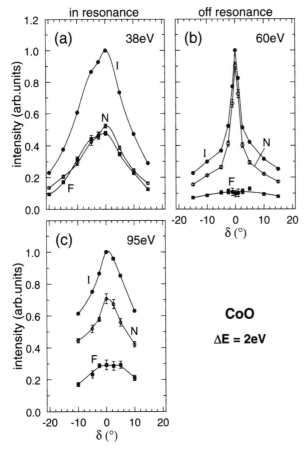

Fig. 5.15. CoO. Scattering-geometry dependence of spin-integrated intensity $I(\Delta E)$ (•), spin-flip intensity $F(\Delta E)$ (■) and nonflip intensity $N(\Delta E)$ (○) at 2 eV energy loss ($^4T_{1g} \rightarrow {}^4A_{2g}$ (^4F) excitation). (**a**), (**c**) in resonance, (**b**) off resonance. δ is the rotation angle of the sample; $\theta_i = 45° + \delta$, $\theta_d = 45° - \delta$

The dramatic intensity decrease at *off-resonant* primary energies (see Fig. 5.14b) is confined to the nonflip part of the spectra only (Figs. 5.15b and 5.16b). The strongly decreasing nonflip intensity is superposed on a slowly varying, nearly isotropically distributed spin-flip intensity. Obviously, a high contribution of dipole-scattering events, which are expected to occur in the nonflip intensity only (Sect. 3.3.2) and to be distributed in a narrow dipolar lobe around the specular scattering geometry (Sect. 3.2.2, (3.8), Figs. 3.1 and 3.3)[8], is superposed on exchange-scattering processes with a wide an-

---

[8] The estimates leading to the very narrow dipolar-lobe shape of the differential scattering cross section in dielectric dipole-scattering theory ((3.8), Fig. 3.3) are strictly valid in the limit of the nearly negligible energy and momentum transfer

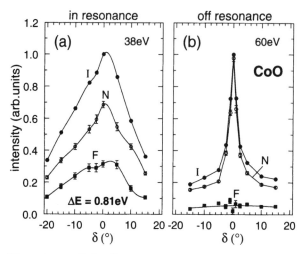

Fig. 5.16. CoO. Scattering-geometry dependence of spin-integrated intensity $I(\Delta E)$ (•), spin-flip intensity $F(\Delta E)$ (■) and nonflip intensity $N(\Delta E)$ (○) at 0.81 eV energy loss ($^4T_{1g} \to {}^4T_{2g}$ (^4F) excitation). (**a**) in resonance, (**b**) off resonance. δ is the rotation angle of the sample, $\theta_i = 45° + \delta$, $\theta_d = 45° - \delta$

gular spread, represented by the spin-flip intensity. As already pointed out in Sect. 3.2.3, exchange scattering, as a special case of impact scattering, is often found to depend weakly on the scattering geometry. For the dominant d–d excitation of NiO (1.58 eV), an equivalent behavior is observed off resonance, as illustrated in Fig. 5.18b for 80 eV primary energy. Here also, a high nonflip intensity, concentrated in a dipolar-lobe-like feature around the specular scattering geometry, is superimposed on a spin-flip intensity which is nearly independent of the scattering geometry.

The prevailing nonflip intensity arising from dipole-scattering events (Figs. 5.15b, 5.16b, 5.18b) is responsible for the nearly identical polarizations of the scattered and incident electrons ($P_s/P_0 \approx 0.8$) observed in the specular scattering geometry for NiO and CoO off resonance (Sect. 5.2.3.2, Fig. 5.13a,b). In accordance with the strong decrease of the nonflip intensity outside the dipolar lobe and the more isotropically distributed exchange scattering processes, resulting in barely varying spin-flip and nonflip intensities in off-specular scattering geometries, the polarization of the scattered electrons P_s decreases strongly if the sample is rotated (Figs. 5.17b and 5.19b).

In resonance, the scattering-geometry dependence of the intensity and polarization of the electrons scattered with 0.81 eV and 2 eV energy loss from

---

that occur in high-energy transmission EELS. But, as pointed out in Sect. 3.2.2, the confinement of the dipole-scattered electrons in a narrow lobe around the specular scattering geometry is often also found in low-energy EELS. This is the case for the d–d excitations here. Even for relatively low primary energies of 60–80 eV, a dipolar-lobe-like distribution of the electrons scattered with 0.8–2 eV energy loss appears, but with a larger FWHM than that predicted by (3.8).

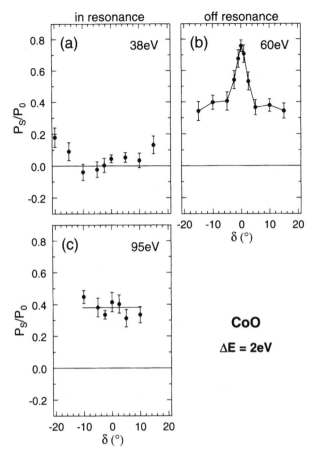

Fig. 5.17. CoO. Scattering-geometry dependence of the normalized polarization of the scattered electrons at 2 eV energy loss ($^{4}T_{1g} \rightarrow {^{4}A_{2g}}$ (^{4}F) excitation). (a), (c) in resonance, (b) off resonance. δ is the rotation angle of the sample, $\theta_i = 45° + \delta$, $\theta_d = 45° - \delta$

CoO and 1.58 eV energy loss from NiO deviates strongly from that observed off resonance (Figs. 5.15–5.19). The spin-integrated, spin-flip, and nonflip intensities decrease slowly and in proportion when the sample is rotated towards off-specular scattering geometries. The spin-flip intensity is high and the ratio of the nonflip to the spin-flip intensity N/F is constant [55]. Obviously, the intensity enhancement in resonance is confined to the exchange-scattering processes with a wide angular spread, exclusively. In contrast to the conditions at off-resonant primary energies, where the relatively weak intensity of scattered electrons mainly arises from direct dipole scattering, the contribution of dipole-scattering processes to the total scattering intensity is negligible in resonance. A dipolar-lobe-like nonflip feature is not observed, and the angular distribution of the scattered electrons is totally determined

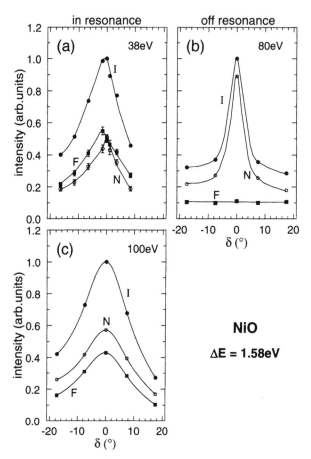

Fig. 5.18. NiO. Scattering-geometry dependence of spin-integrated intensity $I(\Delta E)$ (•), spin-flip intensity $F(\Delta E)$ (■), and nonflip intensity $N(\Delta E)$ (○) at 1.58 eV energy loss ($^3A_{2g} \rightarrow {}^3T_{1g}$, ^1E_g (^3F, ^1D) excitation). (**a**) in resonance, freshly cleaved crystal; (**b**) off resonance, sputtered surface; (**c**) in resonance, sputtered surface; δ is the rotation angle of the sample, $\theta_i = 45° + \delta$, $\theta_d = 45° - \delta$

by exchange scattering. This can be understood from the considerations in Sect. 3.2: excitations by the dipole-scattering mechanism occur far above the target surface (Sect. 3.2.2), where the electron cannot be captured into the target and formation of a compound state is impossible. On the other hand, the probability of exchange processes, which require the close approach of the incoming electron to the target surface (Sect. 3.2.3), can be expected to be strongly enhanced if the electron is temporarily bound to a target atom or ion, forming a resonant compound state (Sect. 3.2.4.2). This is obviously observed here.

Owing to the high spin-flip intensity and the constant ratio of nonflip to flip intensity [55], the scattered electrons are highly depolarized in resonance,

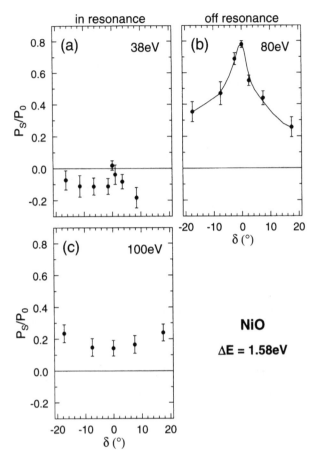

Fig. 5.19. NiO. Scattering-geometry dependence of the normalized polarization of the scattered electrons at 1.58 eV energy loss ($^3A_{2g} \rightarrow {}^3T_{1g}$, ^1E_g (^3F, ^1D) excitation). (**a**) in resonance, freshly cleaved crystal; (**b**) off resonance, sputtered surface; (**c**) in resonance, sputtered surface. δ is the rotation angle of the sample, $\theta_i = 45°+\delta$, $\theta_d = 45° - \delta$

independent of the scattering geometry (Figs. 5.13b and 5.17a,c for CoO, Figs. 5.13a and 5.19a,c for NiO).

The existence of large contributions from dipole-scattering events to the excitation of the 0.81 eV and 2 eV d–d transitions of CoO (Figs. 5.16b and 5.15b), respectively and the 1.58 eV transition of NiO (Fig. 5.18b), shown by the distinct dipolar-lobe-like feature in the nonflip intensity at off-resonant primary energies, indicates nonvanishing dipole matrix elements for these transitions. This leads to the assumption that these transitions are dominantly of the multiplicity-conserving type, which become slightly allowed in the crystal field owing to the relaxation of the parity selection rule (Sect. 2.3.2). In fact, nearly no excitations by nonflip dipole-scattering processes are observed in the case of the multiplicity-changing d–d transitions of MnO (Sect. 5.3.1.3). In multiplicity-changing f–f transitions of Eu^{3+} ions in europiumoxide also, no indications of excitation by dipole-scattering processes are found, as was recently shown [57]. We assign the 0.81 eV energy loss of CoO to the $^4T_{1g} \rightarrow {}^4T_{2g}$ (^4F) transition, in accordance with all other published assignments (Sect. 5.5.3, Table 5.2), and the 2 eV energy loss, which is a subject of controversy in the literature (Sect. 5.5.3, Table 5.2), to the $^4T_{1g} \rightarrow {}^4A_{2g}$ (^4F) excitation. The 1.58 eV energy loss of NiO, which is believed to consist of two close-lying d–d transitions, the $^3A_{2g} \rightarrow {}^3T_{1g}$ (^3F) and the $^3A_{2g} \rightarrow {}^1E_g$ (^1D) excitations (Sect. 5.5.4, Table 5.4), is clearly dominated by the slightly allowed $^3A_{2g} \rightarrow {}^3T_{1g}$ transition, as can be inferred from the considerations above.

Calculations of the impact-scattering cross section [130], valid for off-resonant primary energies, predict the prevalence of nonflip processes ($N/F > 1$) in the excitation of multiplicity-conserving transitions in NiO (Sect. 3.3.3). This is in fact observed in our spin-resolved spectra measured off resonance (Figs. 5.15b, 5.16b, 5.18b): in scattering geometries far off specular, where dipole scattering is negligible and impact and exchange scattering predominate (Sects. 3.2.2, 3.2.3), the nonflip intensity is always found to exceed the spin-flip intensity for the multiplicity-conserving transitions in NiO and CoO investigated.

Recent scattering-geometry-dependent electron energy-loss measurements with unpolarized electrons on NiO/Ni(001) and NiO/Ag(001) films show that the intensity of the 1.58 eV excitation ($^3A_{2g} \rightarrow {}^3T_{1g}$ (^3F), ^1E_g (^1D)) not only is high in the specular scattering geometry, but is also modulated by reciprocal-lattice vectors like the intensity of the elastically scattered electrons – elastically and inelastically (with 1.58 eV energy loss) scattered electrons exhibit similar LEED spots[9] [139]. It was concluded that the dipole-scattering amplitude is high not only in the dipolar lobe (Sect. 3.2.2) around the specular scattering geometry (which corresponds to the (00) reflection), but also near all other LEED spots and therefore reflects the crystal structure. A further weak intensity modulation was found outside the LEED spots. This is attributed to exchange-scattering processes in the excitation of the $^3A_{2g} \rightarrow {}^3T_{1g}$ (^3F), ^1E_g (^1D) transition.

5.3.1.3 Multiplicity-Changing Transitions. In contrast to the behavior of the multiplicity-conserving d–d excitations of NiO and CoO considered above, the scattering-geometry dependence of the multiplicity-changing d–d excitation intensities of MnO differs only slightly in and off resonance. This is illustrated in the 3D plots for MnO (Fig. 5.20) and explicitly shown in spin-resolved measurements of the spin-forbidden $^6A_{1g} \rightarrow {}^4A_{1g}$, ^4E_g (^4G) tran-

---

[9] In these measurements, performed with a 33 eV primary energy and 124° scattering angle ($\theta_i + \theta_d = 56°$), not only the polar incident and detection angles θ_i and θ_d (Sect. 4.1.5) are varied, but also the azimuthal angle. This provides "intensity maps" corresponding to LEED patterns.

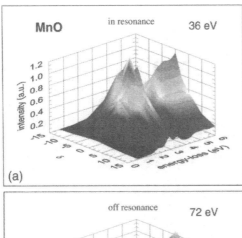

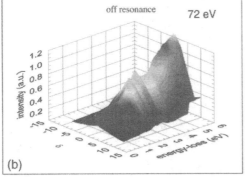

Fig. 5.20 a, b. 3D plots for MnO. The spin-integrated intensity is plotted versus the energy loss and the rotation angle δ of the sample; $\theta_i = 45° + \delta$, $\theta_d = 45° - \delta$

sitions with 2.82 eV excitation energy (Fig. 5.21). A selection of the spectra contributing to Fig. 5.21b is shown in Fig. 5.22.

In resonance (Figs. 5.20a and 5.21a,c), the spin-integrated intensity and the spin-flip and nonflip intensities show a scattering-geometry dependence identical to that observed for the multiplicity-conserving d–d excitations of CoO and NiO (Sect. 5.3.1.2, Figs. 5.15a,c, 5.16a, and 5.18a,c). The scattering process is completely determined by electron exchange in any scattering geometry; the scattered electrons have a wide angular spread and the intensities increase in proportion towards the specular scattering geometry. The ratio of the spin-flip to the nonflip intensity is independent of the scattering geometry [55]. The polarization of the scattered electrons is low and also independent of the scattering geometry. *Off resonance* (Fig. 5.20b and 5.21b,d), the intensities show nearly the same scattering-geometry dependence as in the exchange-dominated resonance case. The spin-flip intensity is also high in the specular scattering geometry. Small-angle nonflip dipole scattering seems to be nearly completely missing, as expected for the spin- and parity-forbidden multiplicity-changing transitions, which remain forbidden even in

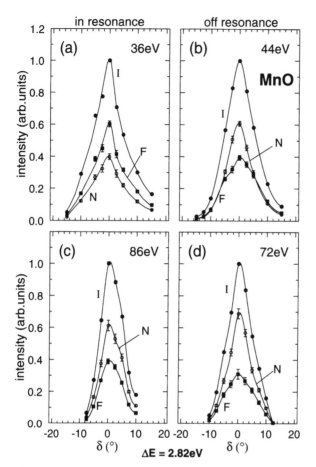

Fig. 5.21. MnO. Scattering-geometry dependence of spin-integrated intensity $I(\Delta E)$ (•), spin-flip intensity $F(\Delta E)$ (■), and nonflip intensity $N(\Delta E)$ (○) at 2.82 eV energy loss ($^6A_{1g} \rightarrow {}^4A_{1g}, {}^4E_g$ (^4G) excitations). (**a**), (**c**) in resonance, (**b**), (**d**) off resonance. δ is the rotation angle of the sample; $\theta_i = 45° + \delta$, $\theta_d = 45° - \delta$

the crystal field of MnO. As a consequence of the high spin-flip intensity and the absence of nonflip dipole scattering at off-resonant primary energies, the polarization of the specularly scattered electrons differs only slightly in and off resonance (Fig. 5.13c). In contrast to NiO and CoO (Fig. 5.13a,b), the polarization in the specular scattering geometry is therefore also low at off-resonant primary energies and deviates strongly from that of the incoming electrons ($P_s/P_0 \ll 1$).

The scattering-geometry dependence of multiplicity-changing, strongly dipole-forbidden septet-quintet f–f transitions of Eu^{3+} ions in europiumoxide has been found recently to be very similar to that of the multiplicity-changing d–d excitations in MnO [57]. For these transitions, the spin-integrated, spin-

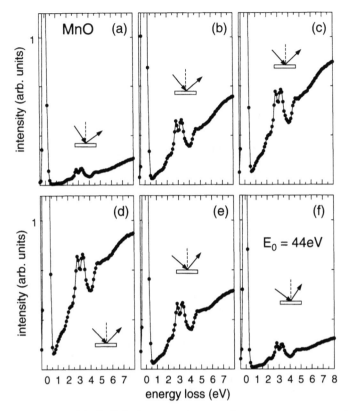

Fig. 5.22. MnO. Spin-integrated energy-loss spectra, measured off resonance with 44 eV primary energy in different scattering geometries. (a) $\delta = -10°$; $\theta_i = 35°$, $\theta_d = 55°$. (b) $\delta = -5°$; $\theta_i = 40°$, $\theta_d = 50°$. (c) $\delta = -2.5°$; $\theta_i = 42.5°$, $\theta_d = 47.5°$. (d) $\delta = 0°$; $\theta_i = \theta_d = 45°$; specular. (e) $\delta = +5°$; $\theta_i = 50°$, $\theta_d = 40°$. (f) $\delta = +10°$; $\theta_i = 55°$, $\theta_d = 35°$

flip, and nonflip intensities also increase in proportion towards the specular scattering geometry – in and off resonance. Also, the polarizations of the scattered and incident electrons differ strongly over the whole primary-energy range between 20 and 160 eV investigated.

Spin-resolved scattering-geometry-dependent measurements of the multiplicity-changing d–d excitations of NiO and CoO have not been performed. The reason is that the multiplicity-changing transitions with low excitation energies (e.g. peak C in Fig. 5.11, see also Tables 5.2 and 5.3) cannot be separated from the very intense multiplicity-conserving transitions. Those with higher excitation energies are superposed on the strong, increasing "background" intensity arising from dipole-allowed transitions across the gap. At off-resonance primary energies ($E_0 < \sim 25$ eV, ~ 45 eV $< E_0 < \sim 90$ eV, and $E_0 > \sim 110$ eV, Figs. 5.5 and 5.13), which are the interesting ones here because the differences in the scattering-geometry dependence between multiplicity-changing and slightly allowed multiplicity-conserving transitions occur off resonance only, all transitions with more than ~2.5 eV excitation energy in NiO and CoO are hardly observable or not observable at all, in particular in the specular scattering geometry (Figs. 5.2 and 5.3). Unlike the situation for the dominant multiplicity-conserving d–d excitations considered above (Sect. 5.3.1.2), it is very difficult here to obtain unambiguous intensity and polarization values, owing to the low signal-to-background ratio.

Recently, Müller et al. [139] investigated the scattering-geometry dependence[10] of the spin-integrated intensity of the 2.7 eV energy-loss peak of NiO ($^3A_{2g} \rightarrow {}^1T_{2g}$ (^1D), $^1A_{1g}$ (^1G) excitations, Sect. 5.5.4, Table 5.3). In the evaluation of their EEL spectra, obtained with unpolarized electrons of 33 eV primary energy and 124° scattering angle (see Sect. 5.3.1.2), the background owing to the increasing intensity of gap transitions was linearly extrapolated and subtracted. The "intensity map" of the 2.7 eV energy-loss peak, obtained by varying the polar as well as the azimuthal angle (Sect. 5.3.1.2), is completely different from that of the $^3A_{2g} \rightarrow {}^3T_{1g}$, ^1E_g excitation (1.58 eV). The intensity is not modulated by reciprocal-lattice vectors as in the case of the $^3A_{2g} \rightarrow {}^3T_{1g}$, ^1E_g excitation (Sect. 5.3.1.2), but seems to reflect the local, fourfold symmetry at the Ni sites. The differences are also attributed here to the different scattering-geometry dependences of exchange and dipole scattering, which are the relevant scattering mechanisms in the excitation of the strongly forbidden $^3A_{2g} \rightarrow {}^1T_{2g}$, $^1A_{1g}$ (2.7 eV) transitions and the slightly allowed, triplet–triplet-dominated $^3A_{2g} \rightarrow {}^3T_{1g}$, ^1E_g (1.58 eV) transition, respectively. Differently from the behavior of the multiplicity-changing $^6A_{1g} \rightarrow {}^4A_{1g}$, ^4E_g (2.82 eV) transitions in MnO measured by us, where the spin-integrated, spin-flip, and nonflip intensities reach their highest values in the specular scattering geometry (Figs. 5.20–5.22), Müller et al. find a relatively small spin-integrated intensity in the specular scattering geometry for the multiplicity-changing transition in NiO investigated. With the chosen experimental conditions and method of evaluation, the intensity is found to increase towards off-specular scattering geometries.

5.3.2 Surface d–d Excitations

5.3.2.1 Introductory Summary. As already illustrated in Fig. 2.8 (see Sect. 2.3.3.1), the transition-metal ions in the bulk and surface experience different crystal fields. The symmetry is reduced at the surface owing to missing oxygen ions; the remaining degeneracy of the 3d states is further lifted, resulting in a further splitting of the 3d states. A variety of states of the surface 3d crystal-field multiplet has been calculated for NiO and CoO assuming a C_{4v}-symmetric crystal field at the surface [44, 73, 181] (Fig. 2.9). Experimentally,

---

[10] For the fixed incident angle of 45° and several scattering angles, they also measured the scattering-*angle* dependence of the 2.7 eV energy-loss peak The results have already been discussed in Sect. 3.2.3.2 (Fig. 3.4).

surface-ion d–d excitations can be identified by comparison of energy-loss spectra obtained from freshly cleaved oxide crystals or freshly prepared oxide films with spectra obtained from sputtered or adsorbate-covered samples (Sect. 2.3.3.1). Before to our investigations reported below, only one d–d transition, requiring 0.56 eV excitation energy, had been identified unambiguously as a surface excitation of NiO ($^3B_{1g} \rightarrow {}^3E_g$ ($^3T_{2g}$), Fig. 2.9, Table 5.4) by means of such comparative measurements [44, 65], [64, p. 103ff.]. For CoO, two surface d–d transitions, of 0.05 eV and 0.45 eV excitation energy, have been found, which are attributed to a $^4A_{2g} \rightarrow {}^4E_g$ ($^4T_{1g}$) and a $^4A_{2g} \rightarrow {}^4B_{2g}$ ($^4T_{2g}$) excitation, respectively [73, 181] (Sect. 5.5.3).

In our scattering-geometry-dependent spin-resolved electron energy-loss measurements, the surface and bulk d–d transitions of NiO are found to show completely different scattering-geometry dependences, providing a possibility to distinguish between them (Sect. 5.3.2.2). By means of such scattering-geometry-dependent measurements, it was possible to identify some weak excitations as surface d–d excitations in NiO. In addition to the well-known $^3B_{1g} \rightarrow {}^3E_g$ ($^3T_{2g}$) transition with 0.56 eV excitation energy, which is the most striking surface d–d transition in energy-loss spectra of freshly cleaved NiO crystals (Fig. 5.23), an energy-loss structure around 2.13 eV was identified as a surface excitation. A further surface d–d excitation has been found, which we assign to the $^3B_{1g} \rightarrow {}^3A_{2g}$ ($^3T_{1g}$) or ^3E_g ($^3T_{1g}$) transition (Fig. 2.9). The measured excitation energy of ≈ 1.33 eV lies very close to the calculated values for those transitions [44] (Table 5.4).

5.3.2.2 Identification of Surface d–d Excitations by Scattering-Geometry-Dependent SPEELS.

Spin-integrated energy-loss spectra of the freshly cleaved NiO surface are shown in Fig. 5.23 for selected scattering geometries, recorded with incident electrons of the resonant primary energy $E_0 = 38$ eV. In going from Fig. 5.23a ($\delta = -17.5°$, $\theta_i = 27.5°$, $\theta_d = 62.5°$) to Fig. 5.23f ($\delta = +8.5°$, $\theta_i = 53.5°$, $\theta_d = 36.5°$), the sample is rotated to more grazing incidence angles and steeper detection angles. Figure 5.23d corresponds to the specular scattering geometry ($\delta = 0°$, $\theta_i = \theta_d = 45°$). In addition to the bulk d–d excitations, which are also visible in spectra obtained from sputtered surfaces (Figs. 5.1a, 5.2, and 5.8), the 0.56 eV surface d–d excitation is clearly observable in any scattering geometry. Owing to changes in the surface stoichiometry caused by sputter-induced defects, the surface d–d excitations arising from the C_{4v}-symmetric crystal field are expected to vanish after sputtering, as mentioned above, and in fact the 0.56 eV excitation is strongly quenched and only weakly indicated in the spectra from sputtered surfaces (Fig. 5.1a).

As can be inferred from Fig. 5.23, the surface and bulk d–d excitations show completely different scattering-geometry dependences. This difference in behavior is clearly visible in the three-dimensional plot in Fig. 5.24, if one concentrates on the 1.58 eV bulk-ion excitations ($^3A_{2g} \rightarrow {}^3T_{1g}$, ^1E_g (^3F, ^1D), Table 5.3, Fig. 2.7) and the 0.56 eV surface-ion excitation ($^3B_{1g} \rightarrow {}^3E_g$

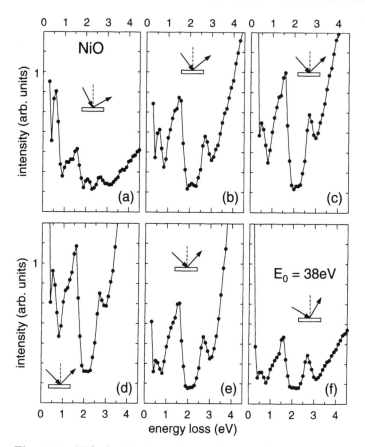

Fig. 5.23. NiO, freshly cleaved. Spin-integrated energy-loss spectra, obtained at 38 eV primary energy in different scattering geometries. (**a**) $\delta = -17.5°$; $\theta_i = 27.5°$, $\theta_d = 62.5°$. (**b**) $\delta = -7.5°$; $\theta_i = 37.5°$, $\theta_d = 52.5°$. (**c**) $\delta = -2.5°$; $\theta_i = 42.5°$, $\theta_d = 47.5°$. (**d**) $\delta = 0°$; $\theta_i = \theta_d = 45°$; specular. (**e**) $\delta = +2.5°$; $\theta_i = 47.5°$, $\theta_d = 42.5°$. (**f**) $\delta = +8.5°$; $\theta_i = 53.5°$, $\theta_d = 36.5°$

($^3T_{2g}$), Table 5.4, Fig. 2.9), which are marked by arrows in Fig. 5.24. Whereas the 1.58 eV bulk excitation shows the intensity with a wide angular spread, increasing slowly and symmetrically towards the specular scattering geometry, which has been found to be typical for bulk d–d transitions of NiO, CoO, and MnO excited by incident electrons at resonance primary energies (Sects. 5.3.1.2, 5.3.1.3), the spin-integrated intensity of the 0.56 eV surface d–d excitation decreases dramatically within the very small angular range of $\pm 2°$ around the specular scattering geometry and increases slowly towards grazing detection angles again. No indication of any further ascent towards more grazing incidence and steeper detection angles ($\delta > 0°$) has been found.

Here, as in the case of the bulk d–d excitations, *spin-resolved* scattering-geometry-dependent measurements of the 0.56 eV energy-loss peak are helpful

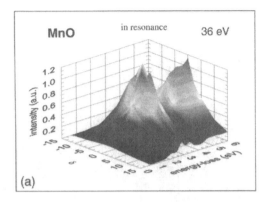

Fig. 5.24. 3D plot for NiO. The spin-integrated intensity is plotted versus the energy loss and the rotation angle δ of the sample; $\theta_i = 45° + \delta$, $\theta_d = 45° - \delta$. The $^3A_{2g} \to {}^3T_{1g}$, ^1E_g (^3F, ^1D) bulk d–d excitations (1.58 eV) and the $^3B_{1g} \to {}^3E_g$ ($^3T_{2g}$) surface d–d excitation (0.56 eV) are marked by *arrows*. Their different scattering-geometry dependences are clearly visible

for clarification (Fig. 5.25a): the high intensity around the specular scattering geometry must be attributed to excitation of the 0.56 eV surface d–d transition by dipole scattering, because this intensity is undoubtedly distributed in a narrow nonflip dipolar lobe, as expected for direct dipole scattering (Sects. 3.2.2, 3.3.2). The dipole-scattering events are superimposed on an intensity, that increases slowly towards off-specular scattering geometries (maximum at $\delta = -17.5°$, $\theta_i = 27.5°$, $\theta_d = 62.5°$), which arises from exchange-scattering processes exclusively, as indicated by the proportional increase of the spin-integrated, spin-flip, and nonflip intensities. In accordance with the high nonflip intensity and low spin-flip intensity within the dipolar lobe, the polarization of the scattered electrons P_s (Fig. 5.25b, filled circles) corresponds to that of the incoming electrons near the specular scattering geometry ($P_s/P_0 \approx 0.8$). P_s is low in off-specular scattering geometries, where the scattering is exchange-dominated and dipole scattering is negligible.

The broadly distributed exchange-assigned intensity of the $^3B_{1g} \to {}^3E_g$ ($^3T_{2g}$) 0.56 eV surface-ion excitation (Fig. 5.25a) behaves very similarly to that of the bulk d–d excitations at resonant primary energies (Figs. 5.15a,c, 5.16a, 5.18a,c, and 5.21a,c), except for the maximum of the exchange-scattering cross section. This is found at grazing detection angles for the surface d–d excitation but in the specular scattering geometry for bulk d–d transitions. The origin of this difference between surface and bulk d–d excitations is not clear at the moment. An enhanced sensitivity for surface excitation processes may play a role in scattering geometries with more grazing detection angles. But an enhancement of the surface sensitivity can also be expected for grazing incidence and this is not observed here. Earlier true scattering-*angle*-dependent EELS measurements with unpolarized electrons

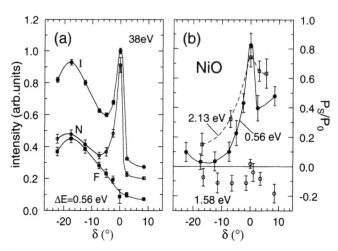

Fig. 5.25. NiO. Scattering-geometry dependence of the energy-loss peaks assigned to surface d–d excitations. The primary energy is 38 eV. δ is the rotation angle of the sample; $\theta_i = 45° + \delta$, $\theta_d = 45° - \delta$. (**a**) Spin-integrated intensity $I(\Delta E)$ (•), nonflip intensity $N(\Delta E)$ (○), and spin-flip intensity $F(\Delta E)$ (■) at 0.56 eV energy loss ($^3B_{1g} \rightarrow {}^3E_g$ ($^3T_{2g}$) surface d–d excitation). The surface excitation is excited by dipole scattering (nonflip dipolar lobe in the specular scattering geometry) and by exchange scattering with a wide angular spread. In contrast to bulk excitations, where the maximum of the exchange-scattering cross section is found in the specular scattering geometry, the maximum is found at grazing detection angles ($\delta = -17.5°$) for the surface excitation. (**b**) Comparison of the normalized polarization of the scattered electrons: 0.56 eV (•) and 2.13 eV (□) surface d–d excitations, 1.58 eV bulk d–d excitation (○) (from Fig. 5.19a)

[65], [64, pp. 108ff.] lead to the assumption that it is really the detection angle which is of importance here. In these measurements, obtained with a fixed incidence angle of 45° and variation of the detection angle,[11] the intensity of the 0.56 eV surface excitation was also found to be enhanced towards grazing detection. In the recent scattering-geometry-dependent energy-loss measurements of Müller et al. [139], which are described in more detail in Sect. 5.3.1.2, the 0.56 eV d–d excitation has also been investigated. Here also an increasing intensity towards grazing detection angles is measured.

As mentioned above, a distinct nonflip dipolar lobe appears in the 0.56 eV surface d–d excitation of NiO, even at 38 eV primary energy, where dipole scattering is negligible for bulk d–d excitations. This may have two possible causes: for the bulk d–d excitations, the resonantly enhanced intensity aris-

---

[11] In scattering-geometry-dependent measurements of the kind that we have used, the scattering angle is fixed and the incidence and detection angles cannot be changed independently (Sect. 4.1.5). For further clarification of the differences in the differential exchange-scattering cross sections of surface and bulk d–d excitations, true scattering-angle-dependent spin-resolved measurements with variation of the scattering angle for a variety of incidence angles are needed.

ing from exchange processes is found around the specular scattering geometry (Sect. 5.3.1.2) – but in the case of the surface excitation, the exchange processes are confined to off-specular scattering geometries (Fig. 5.25a). In the specular scattering geometry they therefore contribute very little to the total scattering intensity; the dipole-scattering events are not superimposed on exchange processes and remain clearly visible. In addition, the probability of excitation by the dipole-scattering mechanism might be enhanced for surface d–d transitions owing to further relaxation of the parity selection rule in the reduced symmetry of the surface crystal field (Sect. 2.3.2).

The weak energy-loss structure at 2.13 eV exhibits an intensity increase towards grazing detection angles corresponding to that of the 0.56 eV surface energy-loss peak (Fig. 5.23). Therefore, the surface-excitation character of this loss peak can be assumed. The peak might be assigned to an excitation into a surface component of a higher 3d multiplet state (Fig. 2.7). In addition to the coincidence of the scattering-geometry dependence of the spin-integrated intensities at 0.56 eV and 2.13 eV energy loss, the scattering-geometry dependence of the polarization at 2.13 eV energy loss also corresponds to that of the 0.56 eV surface d–d excitation (open squares in Fig. 5.25b). But it has to be noted that the polarization values of the 2.13 eV energy-loss peak are not as certain as those of the 0.56 eV energy loss because, especially in the specular scattering geometry, the flanks of the 1.58 eV and 2.69 eV loss peaks and the onset of the dipole-allowed transitions across the optical gap strongly superpose the 2.13 eV excitation, as curve fits show. For the 0.56 eV excitation, on the contrary, superposition with neighboring loss peaks and, especially, the flank of the elastically-scattered-electron peak is found to be negligible (Fig. 5.31), independent of the scattering geometry.

Gorschlüter and Merz [65] attribute the 2.13 eV energy-loss peak to a defect excitation, owing to its weak appearance in optical absorption measurements (which are not surface-sensitive) [148, 166] and in energy-loss spectra obtained from sputtered surfaces [50]. But the 0.56 eV surface excitation is weakly visible for sputtered surfaces, too (Fig. 5.1a), and both excitations, 0.56 eV and 2.13 eV, are much better observable with freshly cleaved crystals. It cannot completely be excluded that the 2.13 eV energy loss arises from a defect, because the excitation energies within the 3d multiplet of a Ni ion adjacent to an oxygen vacancy (Fig. 2.8) differ only slightly from those of excitations within the regular surface 3d multiplet and can hardly be distinguished from them (Sect. 2.3.3.1). Nevertheless, our suggestion of the surface character of the 2.13 eV energy-loss structure is strongly supported by the aforementioned (Sect. 5.3.1.2) scattering-geometry-dependent EELS measurements of Müller et al. [139, 140]. In those measurements, not only are identical scattering-geometry dependences of the 2.13 eV and 0.56 eV excitations found, but also the suggestion of the surface character of the 2.13 eV excitation is additionally substantiated by the fact that these two energy-loss peaks, at 0.56 and 2.13 eV, are the only structures observable for NiO(001) films in the submonolayer range. The bulk d–d excitations are completely missing here.

As can be inferred from the considerations above, the intensities of the energy-loss peaks corresponding to bulk d–d excitations are more or less strongly reduced in off-specular scattering geometries. A similar behavior is observed for the dipole-allowed excitations across the optical gap, especially for the first gap excitation of NiO (the shoulder at ≈ 4.8–$5\,\mathrm{eV}$ energy loss in Figs. 5.1a, 5.7, and 5.33a), which is assigned to the charge-transfer transition [64, p. 70] (Sect. 2.2). The surface d–d excitations, on the contrary, are found to increase again towards grazing detection angles. Therefore, off-specular measurements with grazing detection are helpful for the identification of further surface d–d excitations. In energy-loss spectra where the $0.56\,\mathrm{eV}$ and $2.13\,\mathrm{eV}$ surface energy losses are clearly observable (Fig. 5.23a), the energy-loss peak at $\approx 1.1\,\mathrm{eV}$, which is usually assigned to the $^3A_{2g} \rightarrow {}^3T_{2g}$ (^3F) bulk d–d excitation (Fig. 2.7, Table 5.3), is found to consist of two components at $1.1\,\mathrm{eV}$ and $1.33\,\mathrm{eV}$. This is clearly visible in spectra where the d–d excitations have been fitted by Lorentz profiles (Sect. 5.5.4, Fig. 5.31). We assign the $1.1\,\mathrm{eV}$ peak to a superposition of the $^3A_{2g} \rightarrow {}^3T_{2g}$ (^3F) bulk d–d excitation with the $^3B_1 \rightarrow {}^3B_2$ ($^3T_{2g}$) surface d–d excitation, and the $1.33\,\mathrm{eV}$ peak to the $^3B_1 \rightarrow {}^3A_{2g}$ ($^3T_{1g}$) or ^3E_g ($^3T_{1g}$) surface excitation (Figs. 2.7 and 2.9). The measured excitation energies are very close to the calculated values [44] (Tables 5.3 and 5.4). The participation of surface ions in these excitations can be inferred, because the splitting is clearly visible only in scattering geometries where bulk excitations are reduced and surface excitations are enhanced.

The suggestion of a surface d–d excitation with $1.33\,\mathrm{eV}$ energy loss is also strongly supported by a comparison of energy-loss spectra obtained from sputtered and freshly cleaved crystals under identical experimental conditions (Fig. 5.26). For the sputtered surface (Fig. 5.26a), no indication of a d–d excitation at $1.33\,\mathrm{eV}$ energy loss (dashed line) is found. The bulk d–d excitation peaks at $1.1\,\mathrm{eV}$ and $1.58\,\mathrm{eV}$ (denoted by arrows in Fig. 5.26a) are distinctly separated owing to the absence of the $1.33\,\mathrm{eV}$ surface-ion d–d excitation. In the spectrum from the freshly cleaved crystal, the presence of the $1.33\,\mathrm{eV}$ surface excitation leads to an overlap of the $1.1\,\mathrm{eV}$, $1.33\,\mathrm{eV}$, and $1.58\,\mathrm{eV}$ d–d excitations owing to the limited energy resolution of $230\,\mathrm{meV}$ (Sect. 4.1.1, Table 4.1). Detailed investigations of the scattering-geometry dependence of the intensity and polarization of the $1.33\,\mathrm{eV}$ excitation, for comparison with that of the surface d–d excitations of $0.56\,\mathrm{eV}$ and $2.13\,\mathrm{eV}$ excitation energy (Figs. 5.23–5.25), are impossible owing to this overlap of the energy-loss peaks.

For MnO no surface d–d excitations were observed in our measurements; for CoO one such excitation might be very weakly indicated. The reasons are discussed in Sects. 5.5.2 and 5.5.3 together with the assignment of the energy-loss peaks to the different d–d excitations.

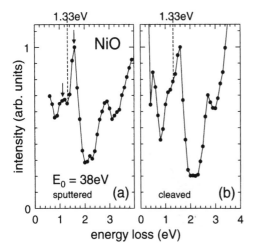

Fig. 5.26. NiO. Spin-integrated energy-loss spectra, measured with 38 eV primary energy in the specular scattering geometry. The energy resolution is 230 meV. (a) sputtered surface (Fig. 5.2); (b) freshly cleaved crystal (Fig. 5.23d). For the sputtered surface, the 1.1 eV and 1.58 eV bulk d–d excitations (*arrows*) are distinctly separated owing to the absence of the 1.33 eV surface d–d excitation. Generally, in spectra from sputtered surfaces, the d–d excitation peaks are found to be slightly broader than in spectra obtained from freshly cleaved crystals. The intensity in the optical gap is slightly enhanced owing to a continuously distributed background arising from excitations into sputter-induced defect states. The intensity of excitations across the optical gap increases less steeply [50]

5.4 Antiferromagnetic Order

5.4.1 Half-Order LEED Spots

The oxide crystals had to be heated to avoid charging (Sect. 4.1.4). Therefore, the NiO crystals were in the antiferromagnetic phase (Néel temperature $T_N = 523$ K, Table 2.1) during the measurements, whereas the CoO and MnO crystals were paramagnetic owing to their lower Néel temperatures of 289 K and 118 K, respectively (Table 2.1). The antiferromagnetic order of the NiO surface was demonstrated by low-energy electron diffraction: at temperatures of 400–450 K, as typically used in the NiO measurements, the samples showed the typical four half-order LEED spots (Fig. 5.27) indicative of a multidomain antiferromagnetic order of the surface [74, 156]. The half-order spots vanished at temperatures above the Néel temperature and appeared again when the sample was cooled down to temperatures lower than T_N.

The half-order spots are attributed to exchange scattering and reflect the magnetic unit cell of the crystal lattice, which is twice as large as the chemical unit cell. Their intensity is very low, ~1–3% of the integer-order beam intensities [156], and their observation is difficult. Therefore, they are often not found [145, 178]. Nevertheless, a number of experimental as well

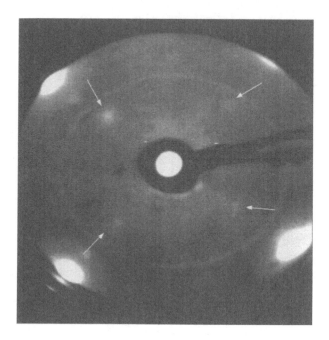

Fig. 5.27. NiO. LEED pattern of a freshly cleaved NiO crystal, obtained with 32 eV electron energy and 7 keV screen voltage [31, 135]. The four weak half-order LEED spots, which are only observable by use of a CCD camera, are indicated by *arrows*. The intense integer-order spots are over-exposed here

as theoretical investigations of the half-order spots and of their behavior, such as the temperature dependence of the intensity, have been published [30, 74, 131, 143, 144, 156, 189, 212]. In our experimental setup, the half-order LEED spots were only observable by use of a CCD camera, where up to 32 images were added directly at the CCD chip.

5.4.2 Influence on d–d Excitations

The primary-energy and scattering-geometry dependences of the energy-loss peaks investigated here and assigned to d–d excitations are nearly identical for antiferromagnetic NiO and paramagnetic CoO (Sects. 5.2 and 5.3). Differences are observed for paramagnetic MnO at off-resonant primary energies only. But these differences must be attributed to the absence of the possibility of excitation by dipole scattering in the strongly dipole-forbidden multiplicity-changing d–d transitions of MnO (Sect. 5.3.1.3) and do not arise from differences in electron-exchange scattering. The electron-exchange processes have been found to show equivalent behavior in the d–d excitations of NiO, CoO, and MnO (Sects. 5.2.1, 5.3.1.2, 5.3.1.3), and electron exchange in the d–d excitations therefore seems to be independent of the magnetic order of the oxides.[12]

Reproducible SPEELS measurements on paramagnetic NiO are impossible because at high temperatures the energy-loss spectra have been found to become altered during the long data acquisition time (Sects. 4.2.1, 5.2.3) needed for spin-resolved measurements. The d–d excitation peaks are reduced and the background intensity in the optical gap increases, which we attribute to changes in stoichiometry during heating. This suggestion is supported by the behavior of the LEED spots (Sect. 5.4.1), which are of worse quality if the sample is kept at high temperature for a long time, and by the experience of other groups. Shen et al. [178] mention the decomposition of NiO just at the Néel temperature, and also Gorschlüter and Merz [65] report changes in the surface stoichiometry caused by heating. In the latter EEL measurements the intensity of the 0.56 eV surface d–d excitation (Sect. 5.3.2) was found to decrease strongly within a few hours if the NiO crystal was heated to a temperature (525 K) which exceeded the Néel temperature only slightly. However, at 380 K, for example, the intensity of the surface d–d excitation remains unchanged over several days. Nevertheless, except for these changes in the intensity of the surface excitation, the electron energy-loss spectra measured with unpolarized electrons below and slightly above the Néel temperature differ only negligibly [65], [64, pp. 80ff.].

Electron energy-loss measurements on antiferromagnetic CoO and MnO are not possible, owing to charging caused by the insufficient conductivity of these crystals at low temperature (Sect. 4.1.4). But optical absorption measurements on CoO and MnO above and below the Néel temperature show only small changes in the CoO spectra (such as slight shifts in the d–d excitation energies of the order of 0.05 eV and a weak further splitting of a few excitation peaks), which are attributed to a tetragonal distortion of the crystal field in the antiferromagnetic phase; no changes are observed for MnO [164].

5.5 Assignment of the d–d Excitation Peaks

5.5.1 Introduction

From the results presented in the preceding sections, it is clear that the determination of the excitation energies necessary for comparison with calculated values and assignment of the energy-loss peaks to particular d–d excitations can be done with resonant spin-polarized electron energy-loss spectroscopy

---

[12] The magnetic order seems to be of negligible influence not only on the exchange processes but on the excitation of d–d transitions in general. This can be concluded from the completely identical behavior of the d–d excitations in antiferromagnetic NiO and paramagnetic CoO, in contrast to the different behavior of the d–d excitations in MnO, which is also paramagnetic.

in an optimal way: at the resonant primary energies, especially in the range 36–38 eV, the intensities of all d–d excitation energy-loss peaks reach their highest values (Fig. 5.2–5.4). At these energies not only the dominant d–d excitations of low excitation energy, located in the optical gaps, but also the weaker d–d excitations, superposed on them or on the strongly increasing excitations across the gap, are clearly observable, especially in the spin-resolved spectra (Figs. 5.10 and 5.11). Off resonance, the weaker d–d excitations are nearly invisible in the spin-integrated spectra (Fig. 5.2–5.4), and measurements of spin-resolved energy-loss spectra like those of Figs. 5.10 and 5.11, covering all d–d excitations, are not possible owing to the very low counting rates.

In addition to resonant spin-resolved measurements, variation of the scattering geometry in spin-polarized electron energy-loss spectroscopy is necessary for the measurement and assignment of the d–d excitations: as already discussed in Sect. 5.3.1.2, the results of the spin-resolved scattering-geometry-dependent measurements on CoO clearly demonstrate the multiplicity-conserving character of the d–d transitions with 0.81 eV and 2 eV energy loss. For NiO, the completely different scattering-geometry dependence of the bulk and surface d–d excitations leads to the identification of further, theoretically predicted surface d–d excitations (Sect. 5.3.2).

5.5.2 MnO

Figure 5.28 shows a magnification of an electron energy-loss spectrum from MnO, obtained with incident electrons at the resonant primary energy of 36 eV (Fig. 5.4). The d–d excitations (labeled A–H in correspondence with Fig. 5.10d) were fitted by the Lorentz profiles shown in the lower part of the figure. The strongly increasing intensity beyond the optical gap was fitted by the flank of a Gaussian curve (broken line in Fig. 5.28). The continuous line through the data points is the result of addition of these fits. Only those d–d excitations which are clearly visible in spin-integrated spectra such as Fig. 5.28 or in the spin-resolved spectra (such as Fig. 5.10) have been fitted. As already pointed out in Sect. 5.2.3.1, the weak d–d excitation at 2.4 eV (B), which is hardly visible in the spin-integrated intensity owing to superposition on the intense d–d excitations at 2.13 eV (A) and 2.82 eV (C), is clearly observed in the polarization and in the spin-flip intensity (Fig. 5.10). Also, the 3.82 eV excitation (E) is much more clearly visible in the spin-flip intensity and in the polarization curve.

Such fits have been performed for more than sixty spectra of different primary energies between 28 eV and 100 eV to determine the exact d–d excitation energies. The average excitation energies are given in Table 5.1, together with calculated values and experimental results from optical absorption spectroscopy and unpolarized EELS. They have already been used in the term scheme of the crystal-field multiplet of MnO in Fig. 2.5. The calculated values presented in the second column (table footnote a) have been taken from

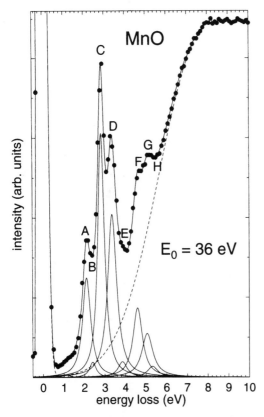

Fig. 5.28. MnO. Spin-integrated energy-loss spectrum, measured in the specular scattering geometry at the resonant primary energy of 36 eV. The *continuous line* is the result of addition of the fits in the *lower part* of the figure. Details are explained in the text

the Orgel diagram of the d^5 configuration, assuming a crystal-field splitting parameter $\Delta_{CF} = 10\,000\,\text{cm}^{-1} \simeq 1.24\,\text{eV}$ for MnO.[13] All measured d–d excitations can be assigned to bulk sextet–quartet excitations from the $^6A_{1g}$ (^6S) ground state of the MnO crystal-field multiplet. Except one d–d excitation of higher excitation energy, all sextet–quartet excitations proposed for the d^5 configuration in the O_h-symmetric crystal field (Tables 2.2 and 2.3) of the bulk transition-metal ion could be measured, some of them for the first time to our knowledge. The assignment of our energy-loss peaks to particular d–d excitations was done by making use of comparison with earlier published val-

---

[13] Most of the published values of the crystal-field splitting of MnO are in the range between $9000\,\text{cm}^{-1} \simeq 1.12\,\text{eV}$ and $11\,000\,\text{cm}^{-1} \simeq 1.36\,\text{eV}$ [28, 85, 86, 94, 164, 180]. They have been measured or calculated from results of optical and x-ray absorption spectroscopy. In the latter case, Δ_{CF} can be directly inferred from the energy difference between the excitations from the O 1s level to the TM $3de_g$ and TM $3dt_{2g}$ states, both mixed with O 2p states [28].

Table 5.1. MnO. Bulk d–d excitation energies energies and their assignment to transitions from the $^6A_{1g}$ (^6S) ground state to quartet final states (Table 2.3, Fig. 2.5). All energies are given in eV. The letters A–H correspond to the notation of Figs. 5.10 and 5.28. The statistical error of the excitation energies determined in the present work is less than 0.01 eV for peaks A, C, D, G and ≈ 0.02 eV for B, E, H

3d-multiplet final state	Calculations		Optical absorption			Energy-loss spectroscopy			Peaks
	a	b	c	d	e	f	g	This work	
$^4T_{1g}$ (^4G)	2.3	2.01	2.1	1.97	2.03	$\approx$2.2	2.2	2.13	A
$^4T_{2g}$ (^4G)	2.9	2.56	2.6	2.51	2.55		} 2.8	2.4	B
$^4A_{1g}$ (^4G)	} 3.3	2.94	} 2.95	2.94	2.97	$\approx$2.8		} 2.82	C
$^4E_{g}$ (^4G)		2.95							
$^4T_{2g}$ (^4D)	3.7	3.5		3.21	3.29	$\approx$3.4	} 3.4	3.31	D
$^4E_{g}$ (^4D)	4.0	3.69		3.46	3.48			3.82	E
$^4T_{1g}$ (^4P)	4.4	4.2					4.6	4.57	F
$^4A_{2g}$ (^4F)	5.4							5.08	G
$^4T_{1g}$ (^4F)	5.5							5.38(?)	H
$^4T_{2g}$ (^4F)	6								

[a] Orgel [154] (with $\Delta_{CF} \approx 1.24$ eV),
[b] van Elp et al. [201],
[c] Pratt and Coelho [164],
[d] Huffman et al. [85],
[e] Iskenderov et al. [94],
[f] Kemp et al. [101],
[g] Jeng and Henrich [98].

ues (Table 5.1). An uncertainty arises in the assignment of the peaks of more than 3.31 eV excitation energy, owing to the missing d–d excitation. In optical absorption spectroscopy [85, 94], a weak absorption structure occurs at 3.4–3.5 eV, which is assigned to the ^4E_g (^4D) excitation (Table 5.1). Also, in most of our spectra an additional shoulder is visible in the spin-flip intensity and in the polarization curve at ≈ 3.5 eV energy loss (see the spin-resolved measurement shown in Fig. 5.10a,b, measured with the 36 eV resonant primary energy), which cannot clearly be separated from the intense $^4T_{2g}$ (^4D) excitation (3.31 eV, D). If this structure arises from a genuine d–d excitation, our assignments of the peaks with $\Delta E > 3.31$ eV must be modified slightly. The question of the existence of a d–d excitation with ≈ 3.5 eV excitation energy could be clarified only by spin-resolved measurements with better energy resolution, which are difficult owing to the low counting rates in SPEELS of dipole-forbidden excitations (Sect. 4.2).

Sextet–doublet excitations are not observed in our spin-polarized electron energy-loss spectra, as expected for these strongly forbidden d–d transitions, which require a change in the spin quantum number of not one, but two

($\Delta S = -2$). No indications of d–d excitations of surface Mn ions are found. This is expected, because the occurrence of distinct surface d–d excitation energy-loss peaks implies an ideal surface stoichiometry (Sect. 2.3.3.1, 5.3.2), but all MnO spectra here were obtained from sputtered surfaces (Sect. 4.2.3), where the surface d–d excitations are generally expected to be strongly quenched (Sects. 2.3.3.1, 5.3.2). However, the prominent 0.56 eV surface d–d excitation of NiO is weakly visible in spectra obtained from sputtered NiO surfaces (Fig. 5.12a), which are of worse quality than those obtained from sputtered MnO. Calculations of surface d–d excitation energies for MnO have not been published up to now, but the excitation energies are expected to lie very close to some of the bulk d–d excitation energies [183]. This reduces drastically the chance of their observation in spectra from sputtered surfaces. Because of the bad quality of freshly cleaved MnO surfaces mentioned earlier (Sect. 4.2.3), it seems doubtful, whether surface d–d excitations of MnO can ever be demonstrated experimentally at all.

At 1.18 eV excitation energy, an additional, weak excitation peak is observed in the MnO spectra (Figs. 5.4, 5.22, and 5.28). The intensity of this loss structure was found to be different in spectra obtained from different surface areas. When the energy-loss peaks which can definitely be assigned to d–d excitations of the Mn^{2+} ions in MnO (Table 5.1) are best observable, the 1.18 eV excitation is hardly visible. It gains in intensity when the spectra are generally of worse quality. We assign the 1.18 eV energy-loss peak to the ^5E_g (^5D) $\rightarrow$ $^3T_{1g}$ (^3H) excitation of Mn^{3+} ions. Mn^{3+} ions have a d^4 configuration, with three t_{2g} electrons but only one e_g electron in the ground state (compare Fig. 2.4a). These ions could be present if the stoichiometry was altered and the oxide Mn_2O_3, which is also stable, was formed instead of MnO, which might be the case in some of the surface facets or at the facets' boundaries in our not really single-crystal MnO samples (Sect. 4.2.3). Previously published energy-loss spectra of Mn_2O_3 show a similar weak energy-loss structure between 1 and 2 eV [101], which is not discussed by the authors. Calculations for different oxidation states of Mn show an enhancement of the crystal-field splitting Δ_{CF} and a reduction of the exchange splitting Δ_{Ex} on going to higher oxidation states [180]. Therefore, the energetically lowest d–d excitation of the Mn^{3+} ion ($t_{2g}^3 e_g \rightarrow t_{2g}^4$) is expected to require a lower excitation energy than the lowest energy in the d^5 configuration ($t_{2g}^3 e_g^2 \rightarrow t_{2g}^4 e_g$; $^6A_{1g}$ (^6S) $\rightarrow$ $^4T_{1g}$ (^4G)), 2.13 eV (Table 5.1), as can be inferred from Fig. 2.4a. Although the real symmetries of the various stable manganese oxides deviate slightly from O_h symmetry, only small polyhedral distortions are expected. Therefore, calculations for octahedral symmetry are epxected to give a good approximation [180], and the Sugano–Tanabe diagram for the d^4 configuration in O_h symmetry [188, p. 110] can be used for a rough estimation of the d–d excitation energies of Mn_2O_3. The exact value of the crystal-field splitting Δ_{CF} for Mn^{3+} ions in Mn_2O_3 is not known, but if we use the published values for Mn^{3+} ions in different crystal-field surroundings

[180], the energetically lowest d–d excitation is found to be the ^5E_g (^5D) →
$^3T_{1g}$ (^3H) transition, requiring an excitation energy of 0.6–1 eV, depending
on the value of Δ_{CF} used. This is in quite good agreement with our measured value of 1.18 eV, taking into account the limited quantitative validity
of the Sugano–Tanabe diagrams (Sect. 2.3.1) and the uncertainty in the actual value of Δ_{CF} in Mn_2O_3. The Mn_2O_3 spectra also exhibit a d–d excitation
at ≈2.4 eV [101]. Therefore, a participation of Mn^{3+} d–d excitations in the
weak energy-loss structure at 2.4 eV (peak B in our spectra, Figs. 5.10 and
5.28, Table 5.1) cannot be excluded.

5.5.3 CoO

As in the case of MnO (Sect. 5.5.2), the d–d excitation energies of CoO were
determined by fitting the corresponding energy-loss peaks by Lorentz profiles.
An example of such fits is given in Fig. 5.29 for a spin-integrated spectrum
measured at the resonant primary energy of 38 eV in the specular scattering
geometry. The increasing intensity of dipole-allowed gap excitations was fitted
by the flank of a Gaussian curve again (the broken line in Fig. 5.29). The
continuous line through the data points is the result of addition of all of
the fits. Only those d–d excitation peaks which are clearly reflected in the
spin-resolved spectra (Fig. 5.11) have been fitted and labeled with capital
letters (A–H): the weak structures C (2.25 eV) and F (3.57 eV) are much
more clearly visible in the spin-resolved spectra, as already mentioned in the
discussion of Fig. 5.11 in Sect. 5.2.3.1.

A broad energy-loss structure appears between 4 and 5 eV excitation energy in the spin-integrated spectra (Fig. 5.29), and it cannot be decided
from these spectra whether this structure should be attributed to a dipole-
allowed transition across the optical gap or to some d–d excitations. From
the spin-resolved spectra (Fig. 5.11), it is clear that this feature consists of
two transitions (G and H), requiring 4.15 eV and 4.6 eV excitation energy.
Both transitions appear as sharp structures in the polarization curve as well
as in the spin-flip intensity and must therefore be attributed to d–d excitations and not to dipole-allowed gap transitions; these would give rise to broad
energy-loss structures, which would be found mainly in the nonflip intensity
(see Fig. 5.12 for NiO).

A variety of energy-loss spectra, obtained with various primary energies
and scattering geometries, have been fitted to determine the d–d excitation
energies as exactly as possible. The average excitation energies obtained from
these fits are given in Table 5.2 together with previously published calculated
and measured values. The calculated excitation energies in the fourth column
(table footnote c) are taken from the Sugano–Tanabe diagram for the d^7
configuration [188, p. 109], using a crystal-field splitting parameter of $\Delta_{CF} =$
1.1 eV [164].

In contrast to MnO (Sect. 5.5.2) and NiO (Sect. 5.5.4), the calculated
value of several d–d excitation energies of CoO differ greatly for the same ex-

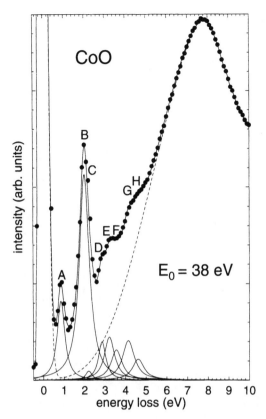

Fig. 5.29. CoO. Spin-integrated energy-loss spectrum, measured in the specular scattering geometry at the resonant primary energy of 38 eV. The *continuous line* is the result of addition of the fits in the *lower part* of the figure. Details are explained in the text

citation. The largest discrepancies occur in the calculations of the crystal-field splitting of the ^{4}F ground state (Fig. 2.6), where the calculated excitation energies of the $^4T_{1g} \rightarrow {}^4A_{2g}$ (^{4}F) transition differ by more than 1 eV (columns 2–4 in Table 5.2). Therefore, the assignment of the measured excitation energies to particular d–d transitions also shows large discrepancies in the literature, and, especially, the assignment of the dominant excitation in the spectra at 2 eV energy loss (Figs. 5.3, 5.11, 5.29, and 5.30) has been a subject of controversy (Table 5.2, columns 5–8): it is assigned to the quartet–doublet transitions $^4T_{1g} \rightarrow {}^2T_{2g}$, $^2T_{1g}$ (^{2}G) as well as to the quartet–quartet transition $^4T_{1g} \rightarrow {}^4A_{2g}$ (^{4}F). Our spin-resolved, scattering-geometry-dependent energy-loss measurements of the 0.81 eV and 2 eV d–d excitations have now definitely solved this problem: it is clear that both energy-loss peaks arise from a slightly allowed quartet–quartet d–d transition, which can be excited by exchange processes as well as direct dipole-scattering processes, resulting

Table 5.2. CoO. Bulk d–d excitation energies and their assignment to transitions from the $^4T_{1g}$ (^4F) ground state to quartet and doublet final states (Table 2.3, Fig. 2.6). All energies are given in eV. The letters A–H correspond to the notation of Figs. 5.11 and 5.29. The assignment of our measured excitation energies mainly follows the calculations of Shi and Staemmler [181]. The statistical error of the excitation energies determined in the present work is less than 0.01 eV for peaks A and B and less than 0.02 eV for the other peaks

3d multiplet final state	Calculations a	b	c	Optical absorption d	e	Energy-loss spectroscopy f	g	This work	Peaks
$^4T_{2g}$ (^4F)	0.8	1.07	1	0.9–1.03	≈0.9	0.85	≈0.9	0.81	A
^2E_g (^2G)	1.69	0.73	1.4	1.61					
$^4A_{2g}$ (^4F)	1.71	3.06	1.9	2.14	≈2	3.2(?)		2.0	B
$^2T_{2g}$ (^2G)	2.36	2.46	} 2.3	2.05			} ≈2	2.25	C
$^2T_{1g}$ (^2G)	2.38	1.92		2.03		2.05			
$^4T_{1g}$ (^4P)	2.56	2.68	2.7	2.26–2.33	≈2.2	2.25			
$^2T_{1g}$ (^2P)	2.82			2.49–2.56				} 2.78	D
$^2A_{1g}$ (^2G)	2.94		3.2	2.6			≈3	} 3.18	E
Crystal-field								3.57	F
terms of		5.6						4.15	G
^2H, ^2F,								4.6	H
^2G, ^2D									

[a]Shi and Staemmler [181], Haßel et al. [73],
[b]van Elp et al. [202],
[c]Sugano et al. [188, p. 109] (with $\Delta_{CF} = 1.1$ eV),
[d]Pratt and Coelho [164],
[e]Kemp et al. [101],
[f]Gorschlüter and Merz [65],
[g]Kämper et al. [100].

in a dipolar-lobe-like intensity, strongly peaked in the specular scattering geometry, which is confined exclusively to the nonflip intensity (Figs. 5.15b and 5.16b). As already mentioned in Sect. 5.3.1.2, we assign the 0.81 eV energy loss to the $^4T_{1g} \to {}^4T_{2g}$ (^4F) transition in accordance with all assignments in the literature (Table 5.2), and the previously controversial 2 eV energy loss to the $^4T_{1g} \to {}^4A_{2g}$ (^4F) excitation. In recently published cluster calculations of the crystal-field multiplet of CoO, the measured 2 eV energy loss is also assigned to the $^4T_{1g} \to {}^4A_{2g}$ (^4F) transition, but unfortunately the calculated value (1.71 eV, second column in Table 5.2) is slightly lower than the experimentally determined value [73, 181]. Nevertheless, except for the $^4T_{1g} \to {}^4A_{2g}$ (^4F) transition, the calculated d–d excitation energies of Shi and Staemmler [181] are in very good accordance with our experimental results. Most of the values taken from the Sugano–Tanabe diagram for $\Delta_{CF} = 1.1$ eV (forth column in Table 5.2) show a similar good agreement with our results.

Here, in particular, the calculated excitation energy of the $^4T_{1g} \rightarrow {}^4A_{2g}$ (^4F) transition (1.9 eV) lies very close to the measured value of 2 eV.

The crystal-field terms of the ^2H, ^2F, ^2G, and ^2D states (Table 2.2 and 2.3) have not been calculated up to now; only one component of ^2D can be inferred from the Sugano–Tanabe diagram [188, p. 109] (fourth column in Table 5.2). Here, three of these terms have been measured for the first time (peaks F–H in Figs. 5.11 and 5.29 and in Table 5.2).

In our spin-polarized electron energy-loss spectra, all measured d–d excitations in CoO must be assigned to d–d transitions of the bulk ions. In contrast to NiO, where the surface d–d excitations are hardly visible in spectra obtained from sputtered surfaces but appear as additional peaks in spectra obtained from freshly cleaved surfaces, especially in off-specular scattering geometries with grazing detection angles (Sect. 5.3.2), no significant differences are observed in the energy-loss spectra of similarly treated CoO surfaces. In-situ cleaved and sputtered crystals provide identical spectra. In addition, all d–d excitation peaks of CoO seem to show a nearly identical scattering-geometry dependence. The intensity is high in the specular scattering geometry and decreases symmetrically when the sample is rotated (Fig. 5.30). No intensity increase towards grazing detection angles, as was found to be typical for the surface d–d excitations of NiO, is observed (Sect. 5.3.2, Figs. 5.23–5.25). Cluster calculations of the crystal-field splitting of the d^7 configuration in the C_{4v} symmetry of the CoO surface predict a variety of surface d–d transitions with excitation energies between ≈ 0.06 and 3.1 eV [73, 181]. From a comparison of high-resolution electron energy-loss spectra obtained from freshly prepared and adsorbate-covered CoO surfaces (Sects. 2.3.3, 5.3.2.1), two weak d–d transitions with excitation energies of 0.05 eV and 0.45 eV have been assigned to the $^4A_{2g} \rightarrow {}^4E_g$ ($^4T_{1g}$) and $^4A_{2g} \rightarrow {}^4B_{2g}$ ($^4T_{2g}$) surface d–d transitions; the calculated values are 0.06 and 0.33 eV [73, 181].

In our spectra a very weak energy-loss structure seems to occur at ≈ 0.5–0.7 eV energy loss in the spin-flip intensity and in the polarization curve (Fig. 5.11). This structure is uncertain and has not been labeled by a capital letter in Fig. 5.11, because it is absolutely invisible in the spin-integrated spectra owing to its superposition on the strongly increasing intensity of elastically scattered electrons and the strong $^4T_{1g} \rightarrow {}^4T_{2g}$ (^4F) transition (0.81 eV), because of the energy resolution of ≈ 230 meV (Sect. 4.1.1, Table 4.1). Perhaps this weak structure indicates the $^4A_{2g} \rightarrow {}^4B_{2g}, {}^4E_g$ ($^4T_{2g}$) d–d transitions of the surface Co^{2+} ions, which are calculated to occur at 0.33 and 0.7 eV energy loss [[181],Hasse:1995]. Indications of other surface d–d transitions with higher excitation energies are found neither in our spin-integrated nor in our spin-resolved energy-loss spectra for CoO (Figs. 5.29 and 5.11, respectively). A d–d excitation of such a low excitation energy as the 0.05 eV ($^4A_{2g} \rightarrow {}^4E_g$ ($^4T_{1g}$)) surface excitation cannot be resolved from the elastically scattered electrons in our measurements, owing to the energy resolution of ≈ 230 meV.

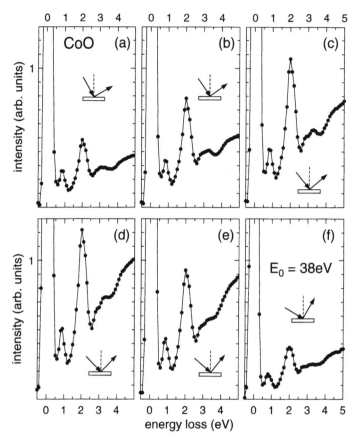

Fig. 5.30. CoO. Spin-integrated energy-loss spectra, obtained at 38 eV primary energy in different scattering geometries. (**a**) $\delta = -15°$; $\theta_i = 30°$, $\theta_d = 60°$. (**b**) $\delta = -10°$; $\theta_i = 35°$, $\theta_d = 55°$. (**c**) $\delta = -5°$; $\theta_i = 40°$, $\theta_d = 50°$. (**d**) $\delta = 0°$; $\theta_i = \theta_d = 45°$; specular. (**e**) $\delta = +5°$; $\theta_i = 50°$, $\theta_d = 40°$. (**f**) $\delta = +15°$; $\theta_i = 60°$, $\theta_d = 30°$

5.5.4 NiO

For freshly cleaved NiO, surface as well as bulk d–d excitations are observed in the energy-loss spectra (Sect. 5.3.2). For an exact determination of all d–d excitation energies in NiO, the energy-loss spectra were fitted according to the procedures used for CoO (Sect. 5.5.2) and MnO (Sect. 5.5.3). As in the case of the bulk d–d excitations, Lorentz profiles were also found to be suitable for reproducing the surface d–d excitation peaks in the energy-loss spectra. An example of such fits is given in Fig. 5.31. This spectrum was measured with incident electrons at the resonant primary energy of 38 eV in an off-specular scattering geometry ($\delta = -17.5°$, $\theta_i = 27.5°$, $\theta_d = 62.5°$; Fig. 5.23a), where the surface d–d excitations are clearly observable.

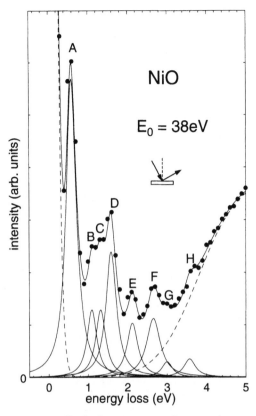

Fig. 5.31. NiO. Spin-integrated energy-loss spectrum, measured in an off-specular scattering geometry ($\delta = -17.5°$; $\theta_i = 27.5°$, $\theta_d = 62.5°$) at the resonant primary energy of 38 eV (see Fig. 5.23a). The *continuous line* is the result of addition of the fits in the *lower part* of the figure. Details are described in the text

The average values of the d–d excitation energies of the NiO bulk and surface, obtained from several curve fits like that shown in Fig. 5.31, are summarized in Table 5.3 (bulk excitations) and Table 5.4 (surface excitations) together with calculated values and other experimental results. The term schemes of the bulk and surface 3d multiplets of NiO (Figs. 2.7 and 2.9) are also based on the excitation energies given in Tables 5.3 and 5.4. The larger statistical error of the weak d–d transitions with 3 eV and 3.55 eV excitation energy (peaks G and H) arises from the fact that these peaks are visible only in a few spectra measured in geometries far off specular. Near the specular scattering geometry, they are not observable, owing to the superimposed high intensity arising from dipole-allowed gap transitions (Fig. 5.23). The assignments of our measured excitation energies to the d–d transitions have been done according to the considerations of Sects. 5.3.1.2 and 5.3.2.

Table 5.3. NiO. Bulk d–d excitation energies and their assignment to transitions from the $^3A_{2g}$ (^3F) ground state to triplet and singlet final states (Table 2.3, Fig. 2.7). All energies are given in eV. The letters B–H correspond to the notation of Fig. 5.31. The statistical error of the excitation energies determined in the present work is less than 0.01 eV for peaks B–F and less than 0.05 eV for G and H

3d multiplet final state	Calculations					Optical absorption		Energy-loss spectroscopy			This work	Peaks
	a	b	c	d	e	f	g	h	i	k		
$^3T_{2g}$ (^3F)	1.13	1.22	1.05	1.0	1.04	1.13	1.08	1.12	1.06	1.1	1.10	B
^1E_g (^1D)	1.92	1.68	1.7		1.74	1.75	1.73	1.6	1.68	1.6	1.58	D
$^3T_{1g}$ (^3F)	1.85	1.98	1.75	1.72	1.75	1.95	1.88	1.7	1.82			
$^1T_{2g}$ (^1D)	2.97	3.0	2.7		2.8	2.75	2.73	2.75		2.7	2.69	F
$^1A_{1g}$ (^1G)	3.06	2.64	2.8		2.86	3.25						G
$^3T_{1g}$ (^3P)	3.21	3.37	3.13		3.12	2.95	2.95	2.9	2.95		3.00	
$^1T_{1g}$ (^1G)	3.56	3.49	3.28		3.49	3.52	3.26	3.1		3.5	3.55	H
$^1T_{2g}$ (^1G)	4.69		4.12		4.39							
^1E_g (^1G)	4.63		4.06		4.42							
$^1A_{1g}$ (^1S)			7.04									

[a] Fujimori and Minami [59],
[b] van Elp et al. [203],
[c] Michiels et al. [Michi 1997a],
[d] Freitag et al. [44], In this paper results of various different calculations are presented. Here the values which show the best agreement with the experimental data are given.
[e] Janssen and Nieuwpoort [96], In this paper results of various different calculations are presented. Here the values which show the best agreement with the experimental data are given.
[f] Newman and Chrenko [148],
[g] Propach et al. [166],
[h] Gorschlüter and Merz [65],
[i] Cox and Williams [24],
[k] Müller et al. [139].

Table 5.4. NiO. Surface d–d excitation energies and their assignment to transitions from the $^3B_{1g}$ (^3F) ground state to triplet final states (Fig. 2.9). All energies are given in eV. The letters A, C, and E correspond to the notation of Fig. 5.31. The statistical error of the excitation energies determined in the present work is less than 0.01 eV

3d multiplet final state	Calculations			Energy-loss spectroscopy			This work	peaks
	a			a	b	c		
^3E_g ($^3T_{2g}$)	0.65	0.54	0.62	0.57	0.6	0.6	0.56	A
$^3B_{2g}$ ($^3T_{2g}$)	1.00	0.86	0.98					
$^3A_{2g}$ ($^3T_{1g}$)	1.30	1.11	1.21	1.62			1.33	C
^3E_g ($^3T_{1g}$)	1.44	1.22	1.38					
(?)						2.1	2.13	E

[a] Freitag et al. [44], various calculations and a measurement,
[b] Gorschlüter and Merz [65],
[c] Müller et al. [139].

Our measurements show that the intense energy-loss peak at 1.58 eV is dominated by the $^3A_{2g} \rightarrow {}^3T_{1g}$ (^3F) transition (Sect. 5.3.1.2). Indications of the $^3A_{2g} \rightarrow {}^3E_g$ (^1D) transition are found neither in scattering-geometry-dependent nor in spin-resolved measurements within the limits of our energy resolution (Sect. 4.1.1). This seems to indicate that the excitation energies differ by less than ≈ 200 meV, as calculated by Michiels et al. [129] and Fujimori and Minami [59] (second and fourth columns in Table 5.3). The long-standing question of whether the $^3T_{1g}$ or the ^1E_g level is higher in energy can be decided only by spin-resolved measurements with very high energy resolution.

5.6 Transitions Across the Optical Gap

5.6.1 Interband and Core-Level Excitations of MnO

Several transitions across the optical gap and transitions from core levels into 3d states have been observed in the energy-loss spectra of NiO, CoO, and MnO [2,60,63,65,86,98,126], [64, p. 68ff.], [200, p. 61], [124,184]. For MnO these excitations, up to 95 eV energy loss, are shown in Fig. 5.32. Owing to the nearly total lack of calculations for the unoccupied states in transition-metal oxides, the assignment of measured excitation energies to particular transitions is not yet clear and there are large differences in the literature (see Table 5.5 for MnO). An exception are the transitions from the 3s and 3p core levels to unoccupied 3d states. For MnO the 3p–3d excitation energies measured in the present work are 47.5 eV and 50.6 eV (peaks e in Fig. 5.32 and Table 5.5). These values are in good agreement with other experimental results [2, 60, 98, 124, 126, 184]. They are nearly identical to the excitation energies of pure Mn, determined from resonant photoemission: the excitation into the ^6D final state of the Mn $3p^5\,3d^6$ configuration requires 48.1 eV excitation energy; the excitation energies for the ^6P and ^4F final states of this configuration are 50.1 eV and 50.7 eV [182]. The measured excitation energy of the broad Mn 3s–Mn 3d transition (around 84 eV, peak f in Fig. 5.32 and Table 5.5) corresponds to the binding energy of the Mn 3s core levels in Mn [76, p. 622] as well as in MnO [200, p. 84ff.] and is in agreement with other EELS studies [184].[14]

As already mentioned in Sect. 2.2, the optical gap of MnO has been calculated to be a mixed form of a charge-transfer and a Mott–Hubbard gap, where the excitation energies of the charge-transfer and Mott–Hubbard transitions ((2.1) and (2.2)) are of comparable size: $\Delta = (7 \pm 1)$ eV, $U =$

---

[14] As in the case of MnO, the excitation energies of the 3s–3d and 3p-3d transitions of CoO and NiO deviate only slightly from the binding energies of the 3s and 3p electrons in the pure metal. This shows that the chemical bond in the transition-metal oxides is of minor influence on the core electrons and that unoccupied 3d states are located close to the Fermi level.

Table 5.5. MnO. Measured excitation energies of core-level and gap transitions (present work) and the assignments made by other authors. The letters a–f correspond to the notation of Figs. 5.32. The Mn 3p–Mn 3d excitation was also measured by McKay et al. [Mcka 1987] and Fujimori et al. [Fuji 1990]

Energy (eV)	[Maye 1996]	[Jeng 1992]	[Huge 1986]	[Stei 1996]	This work	Peaks
7	Mn 3d–Mn 4s		O 2p–Mn 3d Mn 3d–Mn 4s,4p		Charge-transfer Mott–Hubbard	a
10	O 2p–Mn 4s	O 2p–Mn 3d Mn 3d–Mn 4s,4p			O 2p–Mn 4s	b
18	Mn 3d–Mn 4p O 2s–Mn 4p	O 2p–Mn 4s,4p	O 2p–Mn 4p		Mn 3d–Mn 4p	c
28–40	O 2p–O 3s Mn 3d–O 3s	O 2s–Mn 4s,4p			O 2p–O 3p	d
47.5 50.6	Mn 3p–Mn 3d	Mn 3p–Mn 3d		Mn 3p–Mn 3d	Mn 3p–Mn 3d	e
84				Mn 3s–Mn 3d	Mn 3s–Mn 3d	f

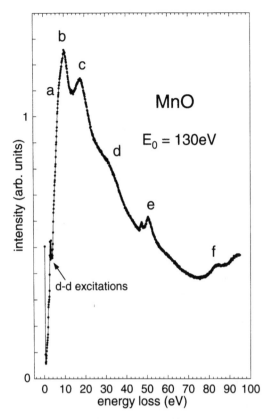

Fig. 5.32. MnO. Spin-integrated energy-loss spectrum, measured in the specular scattering geometry with 130 eV primary energy. The letters a–f mark core-level and gap transitions

(7.5 ± 0.05) eV [60], and $\Delta = 8.8$ eV, $U = 8.5$ eV [201]. In our MnO energy-loss spectra, we assign the first excitation across the optical gap (shoulder a in Fig. 5.32) to this combination of a charge-transfer and a Mott–Hubbard transition. The measured excitation energy (maximum at ≈ 7 eV) is in very good agreement with the calculated values of Δ and U.

From comparisons of photoemission spectra and configuration-interaction calculations [60, 201], the O 2p band is expected to lie at binding energies between approximately 3 and 8 eV. Therefore, transitions from the O 2p band into unoccupied Mn 3d states slightly above the Fermi level may also appear around 7 eV in the energy-loss spectra. But such transitions may also contribute to the broad energy-loss structure around 10 eV (b in Fig. 5.32 and Table 5.5). Photoemission peaks arising from occupied Mn 3d states are found at binding energies between 2 and 4 eV [60, 201]. Transitions from the Mn 3d states into the Mn 4s band, which is expected to lie a few electron volts above the Fermi level, may therefore also contribute to the 7 eV as well as to the 10 eV energy-loss structure. The participation of dipole-forbidden Mn 3d–Mn 4s transitions cannot be excluded in the case of excitation by electrons of low energy, owing to the possibility of electron exchange and electric quadrupole transitions ($\Delta\ell = \pm 2$). Our spin-resolved measurements show that electron exchange plays a minor role. The spin-flip intensity, which indicates the existence of exchange scattering (Sects. 3.3.1, 3.3.2), is small at 7 eV energy loss. Therefore, a high contribution from exchange processes can be excluded. The probability of quadrupole transitions is not known, but the Mn 3s–Mn 3d transition, for example, which violates the same dipole selection rule, is in fact observed in our spin-integrated energy-loss spectra (f in Fig. 5.32). Spin-resolved spectra in the vicinity of the Mn 3s–Mn 3d excitation, which are necessary to decide whether exchange processes contribute to this excitation, have not been measured.

The energy-loss structure around 10 eV (b in Fig. 5.32) is also a subject of controversy in the literature and has been attributed to the O 2p–Mn 4s, the O 2p–Mn 3d, and the Mn 3d–Mn 4s, Mn 4p transitions (Table 5.5). In accordance with the very high intensity of this energy-loss structure, we assign it to a dipole-allowed excitation. From the above energy considerations, this must be the O 2p–Mn 4s excitation, but a contribution of O 2p–Mn 3d or Mn 3d–Mn 4s transitions to this broad energy-loss structure cannot be excluded (see above).

The excitation maximum around 18 eV (c in Fig. 5.32) can be assigned to either the Mn 3d–Mn 4p or the O 2p–Mn 4p excitation for energetic reasons. Owing to the high intensity of this energy-loss structure, we suggest a prevalence of the dipole-allowed Mn 3d–Mn 4p excitation.

The extended, weak energy-loss structure between 28–40 eV (d in Fig. 5.32) occurs at nearly identical energetic positions in the transition-metal oxides. At the primary-electron energy corresponding to this energy loss, the d–d excitations in the various transition-metal oxides exhibit the very strong resonant behavior (see Fig. 5.6 for MnO). As already discussed in Sect. 5.2.1, we attribute this energy-loss structure mainly to the O 2p–O 3p transition. But contributions from other transitions to this very extended energy-loss structure, such as Mn 3d–O 3p transitions or excitations from the O 2s level with ≈ 21 eV binding energy [211], for example, cannot be excluded (Sect. 5.2.1).

5.6.2 Optical Gaps of NiO, CoO, and MnO

The gap widths of the transition-metal oxides can be determined by several experimental methods, such as optical absorption and electron energy-loss spectroscopy; often they are obtained from a comparison of photoemission and bremsstrahlung isochromat spectra. The published values of the gap widths scatter considerably for each of the three oxides (Table 2.1). But, as already mentioned in Sect. 2.2, this is not only a consequence of differences arising from different experimental methods, but also a consequence of different gap definitions. As previously shown [83], [84, p. 188], the gap widths

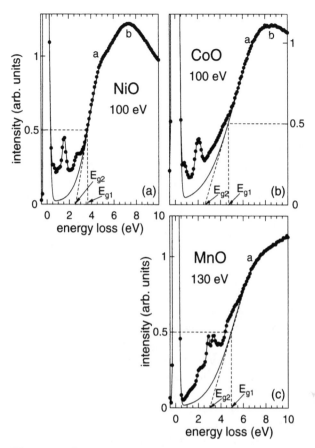

Fig. 5.33. Spin-integrated energy-loss spectra of (a) NiO, (b) CoO, and (c) MnO, measured in specular scattering geometry with high primary energies ($E_0 \geq$ 100 eV). The letters a and b mark the first excitations across the optical gap. For the determination of the gap width by two different methods (E_{g1} and E_{g2}; for details see text), the data have been fitted as described in Sect. 5.5. The Lorentz profiles fitted to the d–d excitations are omitted here for clarity

E_g determined according to different gap definitions from a single optical absorption spectrum of NiO (measured by Powell and Spicer [163])[15] deviate by more than 25%: if the onset of excitation across the optical gap is used, the optical gap of NiO is found to be 3.1 eV; if the first maximum of the absorption spectrum is considered, $E_g = 4.3$ eV. Further gap definitions [83], [84, p. 188], which can also be applied to electron energy-loss spectra, are illustrated for our EEL spectra in Fig. 5.33: in one method, the gap width is taken as that energy where the intensity owing to excitations across the gap

---

[15] More precisely, Powell and Spicer measured the reflectivity and calculated the optical absorption coefficient from these measurements.

reaches half the value of the first maximum of the gap excitations (giving the gap width E_{g1}). A second method is to extrapolate the steep rise towards the first excitation across the gap and define the gap width as that energy where this extrapolation intersects the energy axis (gap width E_{g2}). These methods provide gap widths between the limiting values given by the onset and by the first maximum of the gap transitions. Often the type of gap definition is not given in the literature.

The problems arising in the determination of the gap width become considerably larger when the absorption edge is very smooth, because then the discrepancies between the gap widths provided by the different methods of evaluation of the spectra increase strongly. This is the case for CoO: in optical absorption [163], as well as in electron energy-loss spectroscopy (Fig. 5.33b), the absorption edge beyond the optical gap increases much more smoothly than for NiO. If the gap width is determined from the optical absorption coefficient given by Powell and Spicer [163], according to the procedures described above, the gap width is found to be in the range between 2.5 eV and 6.5 eV.

If the gap width is determined from low-energy EEL spectra, difficulties additional to those for optical absorption spectroscopy arise. Whereas in optical absorption spectroscopy the absorption in the gap owing to the d–d excitations is orders of magnitude lower than for the excitations across the gap (Sect. 2.3.2), the d–d excitation peaks in low-energy EELS are of the same order of magnitude as and only slightly lower than the gap excitations (Sect. 5.1). In resonance, they actually can reach the intensity of the gap excitations (Fig. 5.1). Even at high primary energies of more than 100 eV, where exchange processes are usually expected to be negligible (Sect. 3.2.3.2), the intensity of the d–d transitions remains relatively high in the transition-metal oxides (Fig. 5.33).[16] In particular, MnO and CoO have a high number of d–d excitations owing to the high number of terms in the d^5 and d^7 configurations and their multiple splitting in the O_h symmetric crystal field (Tables 2.2 and 2.3); the ones of higher excitation energy (Tables 5.1 and 5.2) are superposed on the excitations across the optical gap. Therefore, it is impossible to determine definitely the onset of excitation across the optical gap, as can be done in optical absorption spectroscopy.

The absorption edge seems generally to be smoother in energy-loss spectra than in optical absorption spectra. This effect, which cannot fully be attributed to the poorer energy resolution in energy-loss spectroscopy, leads to strong deviations between the gap widths determined by optical absorption spectroscopy and by EELS, if the extrapolation of the increasing intensity beyond the gap is used (E_{g2} in Fig. 5.33).

---

[16] The dominant d–d excitations of NiO and CoO at 1.58 eV and 2 eV energy loss are even weakly visible in energy-loss spectra recorded with 1200 eV primary energy [65], [64, p. 72] (Fig. 5.7).

The first maximum beyond the optical gap is less pronounced in EELS, as a comparison of the spectra of Powell and Spicer [163] and the energy-loss spectra shows: in the optical absorption coefficient of NiO, the first excitation across the gap is indicated by a distinct shoulder at 4.3 eV, appearing on the flank of the 4.9 eV excitation. In EELS a shoulder between 4.8 and 5 eV is observed (a in Fig. 5.33a; see also Fig. 5.7). In the case of CoO the first gap excitation, occurring in the optical absorption spectrum at 6.5 eV, is only very weakly indicated in the EEL spectra. It appears as a hardly visible shoulder at 6.5–7 eV energy loss (a in Fig. 5.33; see also Fig 5.7). However, if the energy where the increasing intensity across the optical gap reaches half the value of the first maximum is taken as the gap width (E_{g1}), the optical absorption and electron energy-loss spectra provide nearly identical gap widths in the case of NiO and CoO: for the determination of the gap width, we used the fitted energy-loss spectra (Sect. 5.5) to separate the d–d excitations from the strongly increasing excitations across the gap. In Fig. 5.33, the continuous line through the data points is the fitted curve, as before. But the Lorentz profiles fitted to the energy-loss peaks assigned to d–d excitations (see Figs. 5.28, 5.29, and 5.31) are omitted for clarity, because only the gap transitions are of interest here. From such fits, a gap width (E_{g1}) of 3.7 eV was obtained for NiO (optical absorption: 3.75 eV [84, p. 188]); the gap width of CoO was found to be 4.7 eV (optical absorption: ≈ 5 eV [163].[17] For MnO, optical absorption spectra of such high quality as those measured by Powell and Spicer [163] for NiO and CoO have not been published. If we determine the gap width from our energy-loss spectra (Fig. 5.33c) by a procedure analogous to that used for CoO and NiO, a gap width of 4.9 eV is obtained.

For CoO, it is not clear at present whether it belongs either to the class of charge-transfer insulators like NiO or to the mixed form of Mott–Hubbard and charge-transfer insulators like MnO, where the Coulomb correlation energy and the charge-transfer energy are of comparable size (Sect. 2.2). If one compares the electron energy-loss spectra of the three oxides (Fig. 5.33), the behavior of the absorption edge of MnO corresponds to that of CoO. For both of these oxides, the intensity of those transitions which determine the optical gap increases very smoothly, whereas NiO exhibits a very steep rise towards the first intensity maximum beyond the optical gap, in absorption spectra as well as in EEL spectra. This suggests that the gap widths of CoO and MnO are determined by the same kind of excitation process, which is different from that in NiO. The observed similar behavior of the first gap transitions in CoO and MnO, unlike that for NiO, might be a hint that CoO

---

[17] The only other published gap width for CoO determined from electron energy-loss spectra is 2.5 eV [64, p. 72]. This in contradiction to our value of 4.7 eV at first sight, but here the different gap definitions come into play again: Gorschlüter determines the gap by extrapolating the increasing intensity across the optical gap (E_{g2} in Fig. 5.33). If we apply the same method to our spectra (Fig. 5.33b) we obtain a corresponding gap width of ≈ 2.6 eV.

is not a pure charge-transfer insulator like NiO, but in fact is a mixed form of Mott–Hubbard/charge-transfer insulator like MnO.

6. Summary

Investigations of the physical and chemical properties of transition-metal oxides have a long history in solid-state and surface physics owing to the wide range of technical applications of these compounds. They have been used in lasers and catalysis for decades; their use in various sensors – e.g. for gases, high pressures, and magnetic fields – followed more recently. Not least a copper oxide is responsible for the occurrence of high-temperature superconductivity in various perovskites. For understanding and optimizing these applications, a detailed knowledge of the electronic structure of the bulk and also the surface of the transition-metal oxides is indispensable.

The "simple" transition-metal monoxides NiO, CoO, and MnO, often thought to be model substances for oxides with a more complicated structure, such as the high-temperature superconductors, do not show metallic conductivity, despite an incompletely filled 3d shell, but instead are insulators with a gap of several electron volts. With the discovery of their insulating nature more than 60 years ago, they were among the first solids found for which a variety of physical properties could not be described by one-particle band-structure calculations. These oxides belong to the class of Mott–Hubbard and charge-transfer insulators, where a strong Coulomb repulsion prevents the formation of a 3d band. The 3d electrons remain localized at the transition-metal ions and cannot move freely through the crystal lattice; an energy of several electron volts is needed for their transfer between neighboring transition-metal ions, leading to the insulating behavior. The localized 3d electrons behave very similarly to electrons in free atoms or ions, but with one significant exception: the degeneracy of the d states related to the magnetic quantum number m_ℓ is partially lifted owing to the crystal field of the six oxygen ions surrounding each transition-metal ion octahedrally – a "crystal-field multiplet" of 3d states arises. Surface transition-metal ions experience a different crystal field owing to missing oxygen ions at the surface. The degeneracy is further lifted – the bulk and surface 3d crystal-field multiplets differ considerably. Generally, excitations within the crystal-field multiplet, the d–d excitations, are dipole forbidden because they violate the parity selection rule $\Delta\ell = \pm 1$. Transitions between initial and final states of identical multiplicity (so-called multiplicity-conserving transitions) become slightly allowed in the crystal field – but the dipole matrix elements remain small and the transitions are only weakly visible in optical absorption spectra. Multiplicity-changing transitions, i.e. those in which the initial and final states have different multiplicities, remain strongly forbidden because they violate the spin selection rule $\Delta S = 0$ additionally and are therefore hardly excitable with light. However, both multiplicity-conserving and multiplicity-changing d–d transitions are easily excited by electrons owing to the possibility of excitation accompanied by electron exchange.

In the work described here, the electronic structures of the transition-metal monoxides NiO, CoO, and MnO have been examined by means of spin-polarized electron energy-loss spectroscopy (SPEELS). This experimental method could be said to have been predestined for such investigations of electronic structures, where optical methods are hardly applicable, owing to the limitations imposed by dipole selection rules. We concentrated mainly on the localized 3d electrons of the bulk and surface transition-metal ions and their dipole-forbidden excitations within the 3d crystal-field multiplet. Some measurements of interband and core-level excitations were also included and the gap widths were determined.

Apart from the determination of the d–d excitation energies, the interaction between the incident electrons and the target, leading to excitation of the target and the corresponding inelastic electron-scattering process, was investigated thoroughly by scattering-geometry-dependent and primary-energy-dependent spin-resolved measurements in the energy range between 20 and 130 eV. Depending on the type of excitation (multiplicity-changing or multiplicity-conserving), the scattering geometry, and the primary energy, different scattering mechanisms were found to contribute to the d–d excitations. Excitations by electron-exchange as well as by the dipole-scattering mechanism were observed. At certain primary energies, strong resonant scattering due to the formation and decay of an intermediate compound state, similar to that observed in electron–atom and electron–molecule scattering, was found. These resonances are of central significance for the determination of the d–d excitation energies. Our results, which allow us to distinguish between surface and bulk d–d excitations and – to a certain degree – between final states of different multiplicity, are briefly summarized here.

Electron exchange is found to be of considerable significance for the excitation of all d–d transitions, even at high primary energies of more than 100 eV, which exceed the d–d excitation energies by two orders of magnitude. This is surprising at first sight, because it is in contrast to the general expectation derived from electron–atom scattering, where exchange is thought to be significant only for primary energies close to the excitation energy. Stimulated by our experimental results, others have recently calculated the scattering cross sections for some d–d excitations in NiO. In these calculations, which use a scattering potential provided by Ni ions embedded in the crystal field of the surrounding oxygen ions in NiO, the significance of exchange in the d–d excitations for energies far above the excitation threshold is also shown.

The d–d excitations exhibit striking resonances in the energy range investigated; a strong resonance, occurring at identical primary energies in the three oxides (36–38 eV), and a weaker one, at higher energies that are different for the three oxides (100–102 eV in NiO, 95 eV in CoO, and 84–85 eV in MnO). The resonant enhancement of the d–d excitations is attributed to the interference of regular d–d excitations, possible at any primary energy, with d–d transitions excited via formation and decay of an intermediate compound state. The formation of the compound state is possible at certain primary energies only, corresponding to inner excitation thresholds. This is the transition-metal 3s–3d excitation in the case of the high-energy resonance and the O 2p–O 3p excitation in the case of the resonance at 36–38 eV. Whereas the primary energy for the 3s–3d resonance is shifted according to the different binding energies of the 3s electrons in the different transition metals, the O 2p–O 3p resonance occurs at nearly identical energies in the three oxides owing to the nearly identical positions of the oxygen levels in the transition-metal oxides.

At off-resonant primary energies, the intensity of the d–d excitations is strongly reduced for all of the three transition-metal oxides. Only the dominant excitations remain clearly visible; the weaker ones are hardly visible or not visible at all. Therefore, a knowledge of the resonant primary energies is essential for the determination of the d–d excitation energies and comparison of the experimental values of these energies with calculated ones. Especially at 36–38 eV primary energy, the d–d excitations are excellently observable. This is impressively demonstrated in the MnO spectra: all d–d excitations of MnO are multiplicity-changing and therefore strongly forbidden by the spin as well as the parity, selection rule. The optical transition probabilities for such excitations have been calculated to be orders of magnitude lower than those for dipole-allowed transitions – but in the energy-loss spectra obtained at the resonant primary energy of 36–38 eV these excitations appear with intensities comparable to that of the dipole-allowed transitions across the optical gap.

Some of the weaker d–d excitations of MnO, as well as those of CoO and NiO, which are superposed on the strongly increasing intensity of dipole-allowed transitions across the optical gap or on very intense d–d excitations, remain barely visible in the spin-integrated intensity, even in resonance. But they are observable in the spin-resolved spectra. They appear clearly in the spin-flip intensity and in the polarization of the scattered electrons, owing to a large contribution of spin-flip exchange processes in these d–d excitations, whereas the dipole-allowed transitions across the optical gap contribute mainly to the nonflip portion of the spectra. By means of spin-resolved measurements in resonance, the excitation energies of nearly all sextet–quartet d–d transitions of MnO could be measured. For CoO, a variety of quartet–doublet transitions of higher excitation energies have been measured for the first time.

In resonance, the excitation of all d–d transitions was found to be completely determined by electron-exchange processes. The intensity of the exchange-scattered electrons has a wide angular spread here, as often observed for impact or exchange scattering. The spin-integrated, spin-flip, and nonflip intensities increase symmetrically and in proportion towards the specular scattering geometry. Off resonance, the exchange scattering, with its wide angular spread, is superposed on strong dipole scattering if multiplicity-conserving transitions, which become slightly allowed in the crystal field, are excited. The dipole-scattered electrons are found to be distributed in a small dipolar lobe around the specular scattering geometry in accordance with the expectations from dipole-scattering theory. The dipolar lobe is confined exclusively to the nonflip intensity, which is also expected because dipole-scattering processes are expected to occur far above the target surface, where exchange is impossible. Excitations by dipole scattering are nearly completely missing in the spin-forbidden, multiplicity-changing d–d transitions of MnO. These excitations, which are not excitable by dipole-scattering processes, because they remain highly forbidden even in the crystal field, are nearly exclusively excited by electron exchange and this applies also at off-resonant primary energies. The scattering-geometry-dependent spin-resolved measurements clearly demonstrate the dominance of the multiplicity-conserving d–d transitions in the NiO and CoO spectra. The intense 2 eV excitation of CoO, which has been a subject of controversy in the literature, must definitely be assigned to a slightly allowed quartet–quartet transition ($^4T_{1g} \rightarrow {}^4A_{2g}\,(^4F)$) owing to the considerable contribution of dipole-scattering processes observed at off-resonant primary energies.

With freshly cleaved NiO crystals, d–d excitations of surface Ni ions are observed. Owing to a different crystal field because of missing oxygen ions in the direction of the surface normal, the crystal-field multiplets of the 3d states of the surface and bulk transition-metal ions, and therefore the d–d excitation energies, are different. As shown by a comparison of spin-resolved scattering-geometry-dependent measurements of the well-known surface d–d excitation of 0.56 eV excitation energy with corresponding measurements of the 1.58 eV bulk d–d excitation, the bulk and surface d–d excitations are found to exhibit a completely different scattering-geometry dependence, providing a possibility to distinguish between them. Taking advantage of this possibility and the high intensity in resonance, further surface d–d excitations of NiO have been measured here for the first time. The measured excitation energies of the $^3B_{1g} \rightarrow {}^3A_{2g}$, $^3E_g\,(^3T_{1g})$ transitions correspond excellently to the calculated values. No surface d–d excitations have been found for freshly cleaved CoO crystals. Owing to a poorer cleavage behavior, all spectra of MnO were obtained from sputtered surfaces, where the appearance of surface d–d excitations is not expected.

The widths of the optical gaps of NiO, CoO, and MnO were determined from our electron energy-loss spectra. The gap width is defined here as that energy at which the strongly increasing intensity of transitions across the optical gap reaches half the value of the first maximum beyond the gap. If this definition is chosen out of the variety of possible gap definitions given in the literature, the gap widths obtained from our NiO and CoO spectra (3.7 eV and 4.7 eV) are in accordance with those obtained from optical absorption spectra by use of the same definition. Applying this definition to MnO, where absorption spectra of comparably high quality have not been published, a gap width of 4.9 eV was obtained. For both CoO and MnO, the intensity of the transitions which determine the optical gap exhibits a similar smooth increase. NiO, on the contrary, exhibits a very steep rise towards the first intensity maximum beyond the gap. This might be a hint that equivalent transitions determine the gap in CoO and MnO, but different transitions do so in NiO. Therefore, CoO might belong to the class of mixed Mott–Hubbard/charge-transfer insulators like MnO and not to the class of pure charge-transfer insulators like NiO.

References

1. D. Adler and J. Feinleib, Electrical and optical properties of narrow-band materials, Phys. Rev. B **2**, 3112 (1970)
2. K. Akimoto, Y. Sakisaka, M. Nishijima, and M. Onchi, Electron energy-loss spectroscopy of UHV-cleaved NiO(100), CoO(100), and UHV-cracked MnO clean surfaces, J. Phys. C: Solid State Phys. **11**, 2535 (1978)
3. G. Albanese, A. Deriu, J.E. Greedan, M.S. Seehra, K. Siratori, and H.P.J. Wijn in *Magnetic Properties of Non-Metallic Inorganic Compounds Based on Transition Elements*, Landolt-Börnstein, New Series, Group III, Vol. 27, Pt. g, ed. by H.P.J. Wijn (Spinger, Berlin, Heidelberg, 1992)
4. M. Alonso and E.J. Finn, *Fundamental University Physics III* (Addison Wesley, Reading, London, 1973), p. 166
5. V.I. Anisimov, J. Zaanen, and O.K. Andersen, Band theory and Mott insulators: Hubbard U instead of Stoner I, Phys. Rev. B **44**, 943 (1991)
6. F. Aryasetiawan, O. Gunnarsson, M. Knupfer, and J. Fink, Local-field effects in NiO and Ni, Phys. Rev. B **50**, 7311 (1994)
7. J.M. Auerhammer and P. Rez, Dipole-forbidden excitations in electron-energy-loss spectroscopy, Phys. Rev. B **40**, 2024 (1989)
8. P. Avouris and J. Demuth, Electron energy loss spectroscopy in the study of surfaces, Ann. Rev. Phys. Chem. **35**, 49 (1984)
9. M. Bäumer, D. Cappus, G. Illing, H. Kuhlenbeck, and H.-J. Freund, Influence of the defects of a thin NiO(100) film on the adsorption of NO, J. Vac. Sci. Technol. A **10**, 2407 (1992)
10. E. Bauer and J. Kolaczkiewicz, Low energy electron exchange excitation of 4f transitions and location of 4f levels in rare earths, phys. stat. sol. (b) **131**, 699 (1985)
11. G. Baum, Polarized electron sources, in *Polarization Phenomena in Nuclear Physics 1980*, Part 2, AIP Conference Proceeding, Vol. 69, ed. by G.G. Ohlson, R.E. Brown, N. Jarmie, W.W. McNaughton and G.M. Hale (American Institute of Physics, New York, 1981), p. 785
12. M. Belkhir and J. Hugel, Origin of the d band separation for the transition metal monoxides, Solid State Commun. **70**, 471 (1989)
13. M. Bender, D. Ehrlich, I.N. Yakovkin, F. Rohr, M. Bäumer, H. Kuhlenbeck, H.-J. Freund, and V. Staemmler, Structural rearrangement and surface magnetism on oxide surfaces: a temperature dependent LEED/EELS study of $Cr_2O_3(111)/Cr(110)$, J. Phys.: Condens. Matt. **7**, 5289 (1995)
14. J.S. Blakemore, Semiconducting and other major properties of gallium arsenide, J. Appl. Phys. **53**, R123 (1982)
15. B.H. Brandow, Electronic structure of Mott insulators, Adv. in Phys. **26**, 651 (1977)
16. L.A. Burkova and V.I. Ochkur, Calculations of inelastic electron scattering from atoms in the second Born approximation, Sov. Phys. JETP **49**, 38 (1979)

17. D. Cappus, C. Xu, D. Ehrlich, B. Dillmann, C.A. Ventrice Jr., K. Al Shamery, H. Kuhlenbeck, and H.-J. Freund, Hydroxyl groups on oxide surfaces: NiO(100), NiO(111) and Cr_2O_3(111), Chem. Phys. **177**, 533 (1993)

18. D. Cappus, M. Menges, C. Xu, D. Ehrlich, B. Dillmann, C.A. Ventrice Jr., J. Libuda, M. Bäumer, S. Wohlrab, F. Winkelmann, H. Kuhlenbeck, and H.-J. Freund, Electronic and geometric structure of adsorbates on oxide surfaces, J. Electron Spectrosc. Relat. Phenom. **68**, 347 (1994)

19. D. Cappus, M. Haßel, E. Neuhaus, M. Heber, F. Rohr, and H.-J. Freund, Polar surfaces of oxides: reactivity and reconstruction, Surf. Sci. **337**, 268 (1995)

20. R.J. Celotta and D.T. Pierce, Sources of Polarized Electrons, Adv. Atom. Mol. Phys. **16**, 101 (1980)

21. H.-h. Chou and H.Y. Fan, Effect of antiferromagnetic transition on the optical-absorption edge in MnO, α-MnS, and CoO, Phys. Rev. B **10**, 901 (1974)

22. F.A. Cotton, *Chemical Applications of Group Theory* (Wiley-Interscience, New York, 1971)

23. P.A. Cox, *Transition Metal Oxides* (Oxford University Press, Oxford, 1992)

24. P.A. Cox and A.A. Williams, The observation of surface optical phonons and low-energy electron transitions in NiO single crystals by electron energy loss spectroscopy, Surf. Sci. **152/153**, 791 (1985)

25. L.C. Davis, Photoemission from transition metals and their compounds, J. Appl. Phys. **59**, R25 (1986)

26. I. Davoli, A. Marcelli, A. Bianconi, M. Tomellini, and M. Fanfoni, Multielectron configurations in the x-ray-absorption near-edge structure of NiO at the oxygen K threshold, Phys. Rev. B **33**, 2979 (1986)

27. J.H. de Boer and E.J.W. Verwey, Semi-conductors with partially and with completely filled 3d-lattice bands, Proc. Phys. Soc. **49** (extra part), 59 (1937)

28. F.M.F. de Groot, M. Grioni, J.C. Fuggle, J. Ghijsen, G.A. Sawatzky, and H. Petersen, Oxygen 1s x-ray-absorption edges of transition-metal oxides, Phys. Rev. B **40**, 5715 (1989)

29. F. Della Valle and S. Modesti, Exchange-excited f–f transitions in the electron-energy-loss spectra of rare-earth metals, Phys. Rev. B **40**, 933 (1989)

30. R.E. De Wames and T. Wolfram, Surface spin waves in antiferromagnetic NiO, Phys. Rev. Lett. **22**, 137 (1969)

31. R, Deußen, *Electron Spectroscopy on the Surface of Solids* (in German) "Elektronenspektroskopie an Festkörperoberflächen", Diploma thesis, University of Düsseldorf, 1995

32. T. Dodt, *Complete, Spin-polarized Electron-Scattering Experiment on Ultrathin Fe Films on $Cu_3Au(001)$* (in German) "Vollständiges, spinpolarisiertes Elektronenstreuexperiment an ultradünnen Fe Filmen auf Cu_3Au(001)", Dissertation, University of Düsseldorf, 1988

33. I.A. Drabkin, L.T. Emel'yanova, R.N. Iskenderov, and Ya.M. Ksendzov, Photoconductivity of single crystals of MnO, Sov. Phys. Sol. State **10**, 2428 (1969)

34. H.-J. Drouhin, C. Hermann, and G. Lampel, Photoemission from activated gallium arsenide. I. Very-high-resolution energy distribution curves, Phys. Rev. B **31**, 3859 (1985)

35. H.-J. Drouhin, C. Hermann, and G. Lampel, Photoemission from activated gallium arsenide. II. Spin polarization versus kinetic energy analysis, Phys. Rev. B **31**, 3872 (1985)

36. F.B. Dunning, L.G. Gray, J.M. Ratliff, F.-C. Tang, X. Zhang, and G.K. Walters, Simple and compact low-energy Mott polarization analyzer, Rev. Sci. Instrum. **58**, 1706 (1987)

37. M.I. D'Yakonov, V.I. Perel', and I.N. Yassievich, Effective mechanism of energy relaxation of hot electrons in p-type semiconductors, Sov. Phys. Semicond. **11**, 801 (1977)

38. D.E. Eastman and J.L. Freeouf, Photoemission partial state densities of overlapping p and d states for NiO, CoO, FeO, MnO, and Cr_2O_3, Phys. Rev. Lett. **34**, 395 (1975)

39. J.S. Escher and H. Schade, Calculated energy distributions of electrons emitted from negative electron affinity GaAs: Cs–O surfaces, J. Appl. Phys. **44**, 5309 (1973)

40. C.S. Feigerle, D.T. Pierce, A. Seiler, and R.J. Celotta, Intense source of monochromatic electrons: Photoemission from GaAs, Appl. Phys. Lett. **44**, 866 (1984)

41. J. Fink, Recent developments in energy-loss spectroscopy, Adv. Electron. Electron Physics **75**, 121 (1989)

42. R.A. Forman, G.J. Piermarini, J.D. Barnett, and S. Block, Pressure measurements made by the utilization of ruby sharp-line luminescence, Science **176**, 284 (1972)

43. J. Franck and G. Hertz, On collisions between electrons and the molecules of mercury vapor and the ionization voltage of the vapor (in German) "Über Zusammenstöße zwischen Elektronen und den Molekülen des Quecksilberdampfes und die Ionisierungsspannung desselben", Ber. DPG **16**, 457 (1914)

44. A. Freitag, V. Staemmler, D. Cappus, C.A. Ventrice Jr., K. Al Shamery, H. Kuhlenbeck, and H.-J. Freund, Electronic surface states of NiO(100), Chem. Phys. Lett. **210**, 10 (1993)

45. H.-J. Freund, B. Dillmann, D. Ehrlich, M. Haßel, R.M. Jaeger, H. Kuhlenbeck, C.A. Ventrice Jr., F. Winkelmann, S. Wohlrab, and C. Xu, Adsorption and reaction of molecules on surfaces of metal–metal oxide systems, J. Mol. Cat. **82**, 143 (1993)

46. H.-J. Freund, Adsorption of gases on solid surfaces, Ber. Bunsenges. Phys. Chem. **99**, 1261 (1995)

47. H.-J. Freund, Metal oxide surfaces: electronic structure and molecular adsorption, phys. stat. sol. (b) 192, 407 (1995)

48. H.-J. Freund, H. Kuhlenbeck, and V. Staemmler, Oxide surfaces, Rep. Prog. Phys. **59**, 283 (1996)

49. B. Fromme, *Emission of Circularly Polarized Light From GaAs and GaAsP Under Impact of Polarized Electrons* (in German) "Emission von zirkular polarisiertem Licht beim Beschuß von GaAs und GaAsP mit polarisierten Elektronen", Dissertation, University of Bielefeld, 1988

50. B. Fromme, M. Schmitt, E. Kisker, A. Gorschlüter, and H. Merz, Spin-flip low-energy electron-exchange scattering in NiO(100), Phys. Rev. B **50**, 1874 (1994)

51. B. Fromme, A. Hylla, C. Koch, E. Kisker, A. Gorschlüter, and H. Merz, Spin-flip low-energy electron exchange scattering in NiO(100), J. Magn. Magn. Mat. **148**, 181 (1995)

52. B. Fromme, Ch. Koch, R. Deussen, and E. Kisker, Resonant Exchange Scattering in Dipole-Forbidden d–d Excitations in NiO(100), Phys. Rev. Lett. **75**, 693 (1995)

53. B. Fromme, M. Möller, Th. Anschütz, C. Bethke, and E. Kisker, Electron-exchange processes in the excitations of NiO(100) surface d state, Phys. Rev. Lett. **77**, 1548 (1996)

54. B. Fromme, C. Bethke, M. Möller, Th. Anschütz, and E. Kisker, Electron-exchange processes in bulk and surface d-d excitations in transition-metal oxides, Vacuum **48**, 225 (1997)

55. B. Fromme, M. Möller, C. Bethke, U. Brunokowski, and E. Kisker, Resonant electron exchange excitations in transition-metal oxides, Phys. Rev. B **57**, 12069 (1998)

56. B. Fromme, U. Brunokowski, and E. Kisker, d–d excitations and interband transitions in MnO: a spin-polarized electron-energy-loss study, Phys. Rev. B **58**, 9783 (1998)

57. B. Fromme, V. Bocatius, and E. Kisker, Electron exchange in the f–f excitations of europiumoxide, Phys. Rev. B, submitted

58. A. Fujimori, F. Minami, and S. Sugano, Multielectron satellites and spin polarization in photoemission from Ni compounds, Phys. Rev. B **29**, 5225 (1984)

59. A. Fujimori and F. Minami, Valence-band photoemission and optical absorption in nickel compounds, Phys. Rev. B **30**, 957 (1984)

60. A. Fujimori, N. Kimizuka, T. Akahane, T. Chiba, S. Kimura, F. Minami, K. Siratori, M. Taniguchi, S. Ogawa, and S. Suga, Electronic structure of MnO, Phys. Rev. B **42**, 7580 (1990)

61. V.S. Gornakov, V.I. Nikitenko, L.H. Bennett, H.J. Brown, M.J. Donahue, W.F. Egelhoff, R.D. McMichael, and A.J. Shapiro, Experimental study of magnetization reversal processes in nonsymmetric spin valve, J. Appl. Phys. **81**, 5215 (1997)

62. A. Gorschlüter, R. Stiller, and H. Merz, Dipole forbidden f–f excitation in ytterbium oxide, Surf. Sci. **251/252**, 272 (1991)

63. A. Gorschlüter and H. Merz, EELS study of single crystalline NiO(100), in *International Conference on the Physics of Transition Metals*, ed. by P.M. Oppeneer and J. Kübler (World Scientific, Singapore, 1993), p. 341.

64. A. Gorschlüter, *Electron-Spectroscopic Investigations on Transition-Metal Oxides with Localized 3d Electrons* (in German) "Elektronenspektroskopische Untersuchung von Übergangsmetalloxiden mit lokalisierten 3d-Elektronen", Dissertation, University of Münster, 1994

65. A. Gorschlüter and H. Merz, Localized d–d excitations in NiO(100) and CoO(100), Phys. Rev. B **49**, 17293 (1994)

66. J.B. Goodenough, *Magnetism and the Chemical Bond*, (Interscience, Wiley, New York, 1963)

67. P. Grünberg, Giant magnetoresistance in magnetic layer structures (in German) "Riesenmagnetowiderstand in magnetischen Schichtstrukturen", Phys. Bl. **51**, 1077 (1995)

68. L.A. Grunes and R.D. Leapman, Optically forbidden excitations of the 3s subshell in the 3d transition metals by inelastic scattering of fast electrons, Phys. Rev. B **22**, 3778 (1980)

69. C. Guillot, Y. Ballu, J. Paigné, J. Lecante, K.P. Jain, P. Thiry, R. Pinchaux, Y. Pétroff, and L.M. Falicov, Resonant photoemission in nickel metal, Phys. Rev. Lett. **39**, 1632 (1977)

70. W. Hanke, A. Muramatsu, and G. Dopf, The Jülich Computer Project: new insights into high-temperature superconductivity (in German) "Das Jülicher Computer-Projekt: Neue Erkenntnisse über die Hochtemperatur-Supraleitung", Phys. Bl. **47**, 1061 (1991)

71. G.F. Hanne, Spin effects in inelastic electron-atom collisions, Phys. Rep. **95**, 95 (1983)

72. M. Hassel and H.-J. Freund, NO on CoO(111)/Co(0001): hydroxyl assisted adsorption, Surf. Sci. **325**, 163 (1995)

73. M. Hassel, H. Kuhlenbeck, H.-J. Freund, S. Shi, A. Freitag, V. Staemmler, S. Lütkehoff, and M. Neumann, Electronic surface states of CoO(100): an electron energy loss study, Chem. Phys. Lett. **240**, 205 (1995)

74. K. Hayakawa, K. Namikawa, and S. Miyake, Exchange reflexions in low energy electron diffraction from antiferromagnetic nickel oxide crystal, J. Phys. Soc. Japan **31**, 1408 (1971)

75. V.E. Henrich and P.A. Cox, *The Surface Science of Metal Oxides* (Cambridge University Press, Cambridge, 1994)

76. M. Henzler and W. Göpel, *Surface Physics of the Solid* (in German) "Oberflächenphysik des Festkörpers" (Teubner, Stuttgart, 1991)

77. W. Ho, R.F. Willis, and W.E. Plummer, Observation of nondipole electron impact vibrational excitations: H on W(100), Phys. Rev. Lett. **40**, 1463 (1978)

78. H. Hopster and D. L. Abraham, Spin-dependent inelastic electron scattering on Ni(110), Phys. Rev. B **40**, 7054 (1989)

79. H. Hopster, Electron depolarization by inelastic exchange scattering from Cr^{3+} magnetic moments, Phys. Rev. B **42**, 2540 (1990)

80. H. Hopster, Spin-polarized electron energy loss spectroscopy study of the initial oxidation of Cr(100), J. Vac. Sci. Technol. A **9**, 1929 (1991)

81. J. Hubbard, Electron correlations in narrow energy bands. II. The degenerate band case, Proc. R. Soc. A **277**, 237 (1964)

82. S. Hüfner, P. Steiner, I. Sander, M. Neumann, and S. Witzel, Photoemission on NiO, Z. Phys. B – Condens. Matt. **83**, 185 (1991)

83. S. Hüfner, P. Steiner, I. Sander, F. Reinert, and H. Schmitt, The optical gap of NiO, Z. Phys. B – Condens. Matt. **86**, 207 (1992)

84. S. Hüfner, Electronic structure of NiO and related 3d-transition-metal compounds, Adv. Phys. **43**, 183 (1994)

85. D.R. Huffman, R.L. Wild, and M. Shinmei, Optical absorption spectra of crystal-field transitions in MnO, J. Chem. Phys. **50**, 4092 (1969)

86. J. Hugel and C. Carabatos, Band structure and optical properties of MnO, Solid State Commun. **60**, 369 (1986)

87. J. Hugel and M. Belkhir, Nature of the NiO absorption edge within a spin polarized band scheme, Solid State Commun. **73**, 159 (1990)

88. H. Hsu, M. Magugumela, B. E. Johnson, F. B. Dunning, and G.K. Walters, Use of spin-polarized electron-energy-loss spectroscopy to investigate dipole and impact scattering from transition-metal surfaces, Phys. Rev. B **55**, 13972 (1997)

89. H. Ibach and D.L. Mills, *Electron Energy Loss Spectroscopy and Surface Vibrations* (Academic Press, New York, 1982)

90. H. Ibach, Electron energy loss spectroscopy with resolution below 1 meV, J. Electron. Spectrosc. Relat. Phenom. **64/65**, 819 (1993)

91. H. Ibach, Electron energy loss spectroscopy: the vibration spectroscopy of surfaces, Surf. Sci. **299/300**, 116 (1994)

92. H. Ibach, M. Balden, and S. Lehwald, Recent advances in electron energy loss spectroscopy of surface vibrations, J. Chem. Soc., Faraday Trans. **92**, 4771 (1996)

93. Y.U. Idzerda, D.M. Lind, D.A. Papaconstantopoulos, G.A. Prinz, B.T. Jonker, and J.J. Krebs, Stoner transitions and spin-selective excitations in bcc cobalt, Phys. Rev. Lett. **61**, 1222 (1988)

94. R.N. Iskenderov, I.A. Drabkin, L.T. Emel'yanova, and Ya.M. Ksendzov, Absorption spectrum of MnO single crystals, Sov. Phys. Sol. State **10**, 2031 (1969)

95. L.W. James and J.L. Moll, Transport properties of GaAs obtained from photoemission measurements, Phys. Rev. **183**, 740 (1969)

96. G.J.M. Janssen and W.C. Nieuwpoort, On the ab-initio calculation of d–d spectra in transition metal compounds: the importance of relaxed charge

transfer states, in *International Journal of Quantum Chemistry, Quantum Chemistry Symposium*, Vol. 22, p. 679 (1988)

97. S.-P. Jeng, R.J. Lad, and V.E. Henrich, Satellite structure in the photoemission spectra of MnO(100), Phys. Rev. B **43**, 11971 (1991)

98. S.-P. Jeng and V.E. Henrich, Interference between 3d→3d electron exchange transitions and interband excitations in MnO(100), Solid State Commun. **82**, 879 (1992)

99. T.S. Jones, Application of electron energy loss spectroscopy in surface science, Vacuum **43**, 177 (1992)

100. K.-P. Kämper, D.L. Abraham, and H. Hopster, Spin-polarized electron-energy-loss spectroscopy on epitaxial fcc Co layers on Cu(001), Phys. Rev. B **45**, 14335 (1992)

101. J.P. Kemp, S.T.P. Davies, and P.A. Cox, High-resolution electron energy loss studies of some transition metal oxides, J. Phys.: Condens. Matter **1**, 5313 (1989)

102. J. Kessler, *Polarized Electrons* (Springer, Berlin, Heidelberg, 1985)

103. J. Kirschner, H.P. Oepen, and H. Ibach, Energy- and spin-analysis of polarized photoelectrons from NEA GaAsP, Appl. Phys. A **30**, 177 (1983)

104. T. Kotani, Exact exchange potential band-structure calculations by the linear muffin-tin orbital-atomic-sphere approximation for Si, Ge, C, and MnO, Phys. Rev. Lett. **74**, 2989 (1995)

105. H. Kuhlenbeck, G. Odörfer, R. Jaeger, G. Illing, M. Menges, T. Mull, H.-J. Freund, M. Pöhlchen, V. Staemmler, S. Witzel, C. Scharfschwerdt, K. Wennemann, T. Liedtke, and M. Neumann, Molecular adsorption on oxide surfaces: electronic structure and orientation of NO on NiO(100)/Ni(100) and on NiO(100) as determined from electron spectroscopies and *ab initio* cluster calculations, Phys. Rev. B **43**, 1969 (1991)

106. H. Kuhlenbeck, C. Xu, B. Dillmann, M. Haßel, B. Adam, D. Ehrlich, S. Wohlrab, H.-J. Freund, U.A. Ditzinger, H. Neddermeyer, M. Neuber, and M. Neumann, Adsorption and reaction on oxide surfaces: CO and CO_2 on Cr_2O_3(111), Ber. Bunsenges. Phys. Chem. **96**, 15 (1992)

107. H. Kuhlenbeck, Electronic excitations on clean and adsorbate-covered oxide surfaces, Appl. Phys. A **59**, 469 (1994)

108. P. Kuiper, G. Kruizinga, J. Ghijsen, G.A. Sawatzky, and H. Verweij, Character of holes in $Li_xNi_{1-x}O$ and their magnetic behavior, Phys. Rev. Lett. **62**, 221 (1989)

109. A. Kuppermann, W.M. Flicker, and O.A. Mosher, Electronic spectroscopy of polyatomic molecules by low-energy variable-angle electron impact, Chem. Rev. **79**, 77 (1979)

110. C.E. Kuyatt, J.A. Simpson, and S.R. Mielczarek, Elastic resonances in electron scattering from He, Ne, Ar, Kr, Xe, and Hg, Phys. Rev. **138**, A385 (1965)

111. C.E. Kuyatt and J.A. Simpson, Electron monochromator design, Rev. Sci. Instrum. **38**, 103 (1967)

112. S. Lacombe, F. Cemic, P. He, H. Dietrich, P. Geng, M. Rocca, and K. Jacobi, Resonant electron scattering of physisorbed O_2 on Ag(111), Surf. Sci. **368**, 38 (1996)

113. R.J. Lad and V.E. Henrich, Electronic structure of MnO studied by angle-resolved and resonant photoemission, Phys. Rev. B **38**, 10860 (1988)

114. G. Lee and S.-J. Oh, Electronic structures of NiO, CoO, and FeO studied by 2p core-level x-ray photoelectron spectroscopy, Phys. Rev. B **43**, 14674 (1991)

115. S. Lehwald, W. Erley und H. Ibach, Vibration spectroscopy on solid surfaces (in German) "Schwingungsspektroskopie an Festkörperoberflächen", in *Grenzflächenforschung und Vakuumphysik* (Kernforschunganlage Jülich GmbH, 1984)

116. T.H. Maiman, Optical and microwave-optical experiments in ruby, Phys. Rev. Lett. **4**, 546 (1960)

117. T.H. Maiman, Stimulated optical radiation in ruby, Nature **187**, 493 (1960)

118. F. Manghi, C. Calandra, and S. Ossicini, Quasiparticle band structure of NiO: the Mott–Hubbard picture regained, Phys. Rev. Lett. **73**, 3129 (1994)

119. R.U. Martinelli and D.G. Fisher, The application of semiconductors with negative electron affinity surfaces to electron emission devices, Proc. IEEE **62**, 339 (1974)

120. S. Massidda, A. Continenza, M. Posternak, and A. Baldereschi, Band-structure picture for MnO reexplored: a model GW calculation, Phys. Rev. Lett. **74**, 2323 (1995)

121. J.A.D. Matthew and Y. Komninos, Transition rates for interatomic Auger Processes, Surf. Sci. **53**, 716 (1975)

122. J.A.D. Matthew, G. Strasser, and F.P. Netzer, Multiplet effects and breakdown of dipole selection rules in the $3d$–$4f$ core-electron-energy-loss spectra of La, Ce, and Gd, Phys. Rev. B **27**, 5839 (1983)

123. J.A.D. Matthew, W.A. Henle, M.G. Ramsey, and F.P. Netzer, $4f^7$–$4f^7$ transitions in Gd, oxidized Gd, and epitaxial Gd silicide, Phys. Rev. B **43**, 4897 (1991)

124. B. Mayer, St. Uhlenbrock, and M. Neumann, XPS satellites in transition metal oxides, J. Electron Spectrosc. Relat. Phenom. **81**, 63 (1996)

125. D.S. McClure, Comparison of the crystal fields and optical spectra of Cr_2O_3 and ruby, J. Chem. Phys. **38**, 2289 (1963)

126. J.M. McKay, M.H. Mohamed, and V.E. Henrich, Localized $3p$ excitations in $3d$ transition-metal-series spectroscopy, Phys. Rev. B **35**, 4304 (1987)

127. J.L. McNatt, Electroreflectance study of NiO, Phys. Rev. Lett. **23**, 915 (1969)

128. J.A. Mejias, V. Staemmler, and H.-J. Freund, Electronic states of the $Cr_2O_3(0001)$ surface from *ab initio* embedded cluster calculations, J. Phys.: Condens. Matt. **11**, 7881 (1999)

129. J.J.M. Michiels, J.E. Inglesfield, C.J. Noble, V.M. Burke, and P.G. Burke, Atomic theory of electron energy loss from transition metal oxides, Phys. Rev. Lett. **78**, 2851 (1997)

130. J.J.M. Michiels, J.E. Inglesfield, C.J. Noble, V.M. Burke, and P.G. Burke, Spin and symmetry in low-energy electron energy-loss spectroscopy of transition metal oxides, J. Phys.: Condens. Matt. **9**, L543 (1997)

131. D.L. Mills, Surface effects in magnetic crystals near the ordering temperature, Phys. Rev. B **3**, 3887 (1971)

132. D.L. Mills, Electron energy loss by electron–hole excitations in ferromagnets: the near-specular geometry, Phys. Rev. B **34**, 6099 (1986)

133. S. Modesti, G. Paolucci, and E. Tosatti, f–f excitations by resonant electron-exchange collisions in rare-earth metals, Phys. Rev. Lett. **55**, 2995 (1985)

134. S. Modesti, F. Della Valle, and G. Paolucci, Spin-exchange processes in inelastic electron scattering from metals, Physica Scripta **T19**, 419 (1987)

135. M. Möller, *Complete Spin-Resolved Electron Energy-Loss Spectroscopy on the Transition-Metal Oxides NiO, CoO, and MnO* (in German) "Komplett spinaufgelöste Elektronen-Energieverlustspektroskopie an den Übergangsmetalloxiden NiO, CoO und MnO", Dissertation, University of Düsseldorf, 1999

136. H.R. Moser, B. Delley, W.D. Schneider, and Y. Baer, Characterization of f electrons in light lanthanide and actinide metals by electron-energy-loss and x-ray photoelectron spectroscopy, Phys. Rev. B. **29**, 2947 (1984)

137. N.F. Mott and R. Peierls, Discussion of the paper by de Boer and Verwey, Proc. Phys. Soc. **49** (extra part), 72 (1937)

138. N.F. Mott, The basis of the electron theory of metals, with special reference to the transition metals, Proc. Phys. Soc. A **62**, 416 (1949)

139. F. Müller, P. Steiner, Th. Straub, D. Reinicke, S. Palm, R. de Masi, and S. Hüfner, Full hemispherical intensity maps of crystal field transitions in NiO(001) by angular resolved electron energy loss spectroscopy, Surf. Sci. **442**, 485 (1999)

140. F. Müller, R. de Masi, P. Steiner, D. Reinicke, M. Stadtfeld, and S. Hüfner, EELS investigation of thin epitaxial NiO/Ag(001) films: surface states in the multilayer, monolayer and submonolayer range, Surf. Sci. **459(1–2)**, 161 (2000)

141. S. Nakai, T. Mitsuishi, H. Sugawara, H. Maezawa, T. Matsukawa, S. Mitani, K. Yamasaki, and T. Fujikawa, Oxygen K x-ray-absorption near-edge structure of alkaline-earth-metal and 3d-transition-metal oxides, Phys. Rev. B **36**, 9241 (1987)

142. K. Nakamoto, Y. Kawato, Y. Suzuki, Y. Hamakawa, T. Kawabe, K. Fujimoto, M. Fuyama, and Y. Sugita, Design and read performance of GMR heads with NiO, IEEE Trans. Magn. **32**, 3374 (1996)

143. K. Namikawa, K. Hayakawa, and S. Miyake, Intensity behavior of exchange reflexions in LEED from an antiferromagnetic crystal, J. Phys. Soc. Japan **37**, 733 (1974)

144. K. Namikawa, LEED investigations on temperature dependence of sublattice magnetization of NiO (001) surface layers, J. Phys. Soc. Japan **44**, 165 (1978)

145. F.P. Netzer and M. Prutton, LEED and electron spectroscopic observations on NiO (100), J. Phys. C: Solid State Phys. **8**, 2401 (1975)

146. F.P. Netzer, G. Strasser, and J.A.D. Matthew, Selection rules in electron-excited $4d \rightarrow 4f$ transitions at intermediate incident energies, Phys. Rev. Lett. **51**, 211 (1983)

147. F.P. Netzer and J.A.D. Matthew, Inelastic electron scattering measurements, in *Handbook on the Physics and Chemistry of Rare Earth Metals*, Vol. 10, ed. by K.A. Gschneidner Jr., L. Eyring and S. Hüfner (North-Holland, Amsterdam, 1987), p. 547

148. R. Newman and R.M. Chrenko, Optical properties of nickel oxide, Phys. Rev. **114**, 1507 (1959)

149. N. Nücker, J. Fink, J.C. Fuggle, P.J. Durham, and W.M. Temmerman, Evidence for holes on oxygen sites in the high-T_C superconductors $La_{2-x}Sr_xCuO_4$ and $YBa_2Cu_3O_{7-y}$, Phys. Rev. B **37**, 5158 (1988)

150. H. Ogasawara and A. Kotani, Theory of XAS and EELS in unfilled shell systems, Tech. Rep. ISSP, Ser. A **3050**, 1 (1995)

151. T. Oguchi, K. Terakura, and A.R. Williams, Band theory of the magnetic interaction in MnO, MnS, and NiO, Phys. Rev. B **28**, 6443 (1983)

152. T. Oguchi, K. Terakura, and A.R. Williams, Transition-metal monoxides: itinerant versus localized picture of superexchange, J. Appl. Phys. **55**, 2318 (1984)

153. K. Okada and A. Kotani, Complementary Roles of Co 2p X-ray absorption and photoemission spectra in CoO, J. Phys. Soc. Japan. **61**, 449 (1992)

154. L.E. Orgel, Spectra of transition-metal complexes, J. Chem. Phys. **23**, 1004 (1955)

155. H.K. Paetzold, On the influence of temperature and pressure on electron terms (in German) "Über den Temperatur- und Druckeinfluß auf Elektronenterme in Kristallen", Z. Phys. **129**, 123 (1951)

156. P.W. Palmberg, R.E. DeWames, and L.A. Vredevoe, Direct observation of coherent exchange scattering by low-energy electron diffraction from antiferromagnetic NiO, Phys. Rev. Lett. **21**, 682 (1968)

157. R.E. Palmer and P.J. Rous, Resonances in electron scattering by molecules on surfaces, Rev. Mod. Phys. **64**, 383 (1992)

158. D.T. Pierce, F. Meier, and D. Zürcher, Negative electron affinity GaAs: a new source of spin-polarized electrons, Appl. Phys. Lett. **26**, 670 (1975)

159. D.T. Pierce and F. Meier, Photoemission of spin-polarized electrons from GaAs, Phys. Rev. B **13**, 5484 (1976)

160. D.T. Pierce, R.J. Celotta, G.-C. Wang, W.N. Unertl, A. Galejs, C.E. Kuyatt, and S.R. Mielczarek, GaAs spin polarized electron source, Rev. Sci. Instrum. **51**, 478 (1980)

161. G.E. Pikus and A.N. Titkov, Spin relaxation under optical orientation in semiconductors, in *Optical Orientation*, ed. by F. Meier and B.P. Zakharchenya (Elsevier Science, Amsterdam, 1984), p. 73

162. S.J. Porter, J.A.D. Matthew, and R.J. Leggott, Inelastic exchange scattering in electron-energy-loss spectroscopy: localized excitations in transition-metal and rare-earth systems, Phys. Rev. B **50**, 2638 (1994)

163. R.J. Powell and W.E. Spicer, Optical Properties of NiO and CoO, Phys. Rev. B **2**, 2182 (1970)

164. G.W. Pratt Jr. and R. Coelho, Optical absorption of CoO and MnO above and below the Néel temperature, Phys. Rev. **116**, 281 (1959)

165. M.W.J. Prins, K.-O. Grosse-Holz, G. Müller, J.F.M. Cillessen, J.B. Giesbers, R.P. Weening, and R.M. Wolf, A ferroelectric transparent thin-film transistor, Appl. Phys. Lett. **68**, 3650 (1996)

166. V. Propach, D. Reinen, H. Drenkhahn, and H. Müller-Buschbaum, On the color of "NiO" (in German) "Über die Farbe von 'NiO'", Z. Naturforsch. **33b**, 619 (1978)

167. H. Raether, *Excitations of Plasmons and Interband Transitions by Electrons* (Spinger, Berlin, Heidelberg, 1980)

168. D. Reinen, Color and constitution of anorganic solids VIII B. Light absorption of divalent nickel in the mixed crystals $Mg_{1-x}Ni_xO$ and in tetrahedral coordination (in German) "Farbe und Konstitution bei anorganischen Feststoffen. VIII B. Die Lichtabsorption des zweiwertigen Nickels in den Mischkristallen $Mg_{1-x}Ni_xO$ und in tetraedrischer Koordination", Ber. Bunsenges. **69**, 82 (1965)

169. F. Rohr, K. Wirth, J. Libuda, D. Cappus, M. Bäumer, and H.-J. Freund, Hydroxyl driven reconstruction of the polar NiO(111) surface, Surf. Sci. **315**, L977 (1994)

170. G.A. Sawatzky and J.W. Allen, Magnitude and origin of the band gap in NiO, Phys. Rev. Lett. **53**, 2339 (1984)

171. G.A. Sawatzky, Electronic structure of transition metal compounds as studied by high energy spectroscopies, in *Core-Level Spectroscopy in Condensed Systems*, Springer Series in Solid-State Sciences, Vol. 81, ed. by J. Kanamori and A. Kotani (Springer, Berlin, Heidelberg, 1988), p. 99

172. J.J. Scheer and J. van Laar, Fermi level stabilization at cesiated semiconductor surfaces, Solid State Commun. **5**, 303 (1967)

173. K.-D. Schierbaum, Anorganic and organic "model"-surfaces for chemical sensors (in German) "Anorganische und organische 'Modell'-Oberflächen für chemische Sensoren", Ber. Bunsenges. Phys. Chem. **99**, 1230 (1995)

174. K.D. Schierbaum and W. Göpel, Chemical sensors based upon catalytic re-actions, in *Handbook of Heterogeneous Catalysis*, Vol. 3, Part 8.1, ed. by G. Ertl, W. Knözinger, J. Weitkamp (VCH, Weinheim, 1997), p. 1283

175. M. Schönnenbeck, D. Cappus, J. Klinkmann, H.-J. Freund, L.G.M. Petterson, and P.S. Bagus, Adsorption of CO and NO on NiO and CoO: a comparison, Surf. Sci. **347**, 337 (1996)

176. M.P. Seah and W.A. Dench, Quantitative electron spectroscopy of surfaces: a standard data base for electron inelastic mean free paths in solids, Surf. Interface Anal. **1**, 2 (1979)

177. Z.-X. Shen, C.K. Shih, O. Jepsen, W.E. Spicer, I. Lindau, and J.W. Allen, Aspects of the correlation effects, antiferromagnetic order, and translational symmetry of the electronic structure of NiO and CoO, Phys. Rev. Lett. **64**, 2442 (1990)

178. Z.-X. Shen, J.W. Allen, P.A.P. Lindberg, D.S. Dessau, B.O. Wells, A. Borg, W. Ellis, J.S. Kang, S.-J. Oh, I. Lindau, and W.E. Spicer, Photoemission study of CoO, Phys. Rev. B **42**, 1817 (1990)

179. Z.-X. Shen, R.S. List, D.S. Dessau, B.O. Wells, O. Jepsen, A.J. Arko, R. Bartlet, C.K. Shih, F. Parmigiani, J.C. Huang, and P.A.P. Lindberg, Electronic structure of NiO: correlation and band effects, Phys. Rev. B **44**, 3604 (1991)

180. D.M. Sherman, The electronic structures of manganese oxide minerals, Am. Mineralogist **69**, 788 (1984)

181. S. Shi and V. Staemmler, An ab initio study of local d-d excitations in bulk CoO, at the CoO(100) surface, and in octahedral Co^{2+} complexes, Phys. Rev. B **52**, 12345 (1995)

182. B. Sonntag and P. Zimmermann, XUV spectroscopy of metal atoms, Rep. Prog. Phys. **55**, 911 (1992)

183. V. Staemmler, private communication, 1998/2000

184. P. Steiner, R. Zimmermann, F. Reinert, Th. Engel, and S. Hüfner, 3s- and 3p-core level excitations in 3d-transition metal oxides from electron-energy-loss spectroscopy, Z. Phys. B **99**, 479 (1996)

185. G. Strasser, F.P. Netzer and J.A.D. Matthew, Breakdown of the one electron approximation for $4p$ core electron energy loss spectra in the rare earths, Solid State Commun. **49**, 817 (1984)

186. G. Strasser, G. Rosina, J.A.D. Matthew, and F.P. Netzer, Core level excitations in the lanthanides by electron energy loss spectroscopy. I: 4d and 3d excitations, J. Phys. F: Met. Phys. **15**, 739 (1985)

187. K. Sturm, Electron energy-loss spectroscopy (in German) Elektronen-Energie-Verlust-Spektroskopie, in *Lecture Notes of the XX. IFF-Workshop* (Kernforschungsanlage Jülich GmbH, Jülich, 1980)

188. S. Sugano, Y. Tanabe, and H. Kamimura, *Multiplets of Transition-Metal Ions in Crystals* (Academic Press, New York, 1970)

189. T. Suzuki, N. Hirota, H. Tanaka, and H. Watanabe, Exchange scattering of low-energy electrons from antiferromagnetic NiO, J. Phys. Soc. Japan **30**, 888 (1971)

190. A. Svane and O. Gunnarsson, Transition-metal oxides in the self-interaction – corrected density-functional formalism, Phys. Rev. Lett. **65**, 1148 (1990)

191. Z. Szotek, T.W. Temmerman, and H. Winter, Application of the self-interaction correction to transition-metal oxides, Phys. Rev. B **47**, 4029 (1993)

192. M. Takahashi and J. Igarashi, Electronic excitations in NiO, Ann. Physik **5**, 247 (1996)

193. M. Takahashi and J. Igarashi, Local approach to electronic excitations in MnO, FeO, CoO, and NiO, Phys. Rev. B **54**, 13566 (1996)

194. K. Terakura, A.R. Williams, T. Oguchi, and J. Kübler, Transition-metal monoxides: band or Mott insulators, Phys. Rev. Lett. **52**, 1830 (1984)

195. K. Terakura, T. Oguchi, A.R. Williams, and J. Kübler, Band theory of insulating transition-metal monoxides: band-structure calculations, Phys. Rev. B **30**, 4734 (1984)

196. M.R. Thuler, R.L. Benbow, and Z. Hurych, Photoemission intensities at the $3p$ threshold resonance of NiO and Ni, Phys. Rev. B **27**, 2082 (1983)

197. A.N. Titkov, E.I. Chalkina, É.M. Komova, and N.G. Ermakova, Low-temperature luminescence of degenerate p-type crystals of direct-gap semiconductors, Sov. Phys. Semicond. **15**, 198 (1981)

198. L.H. Tjeng, C.T. Chen, J. Ghijsen, P. Rudolf, and F. Sette, Giant Cu $2p$ Resonances in CuO valence-band photoemission, Phys. Rev. Lett. **67**, 501 (1991)

199. M.D. Towler, N.L. Allan, N.M. Harrison, V.R. Saunders, W.C. Mackrodt, and E. Aprà, *Ab initio* study of MnO and NiO, Phys. Rev. B **50**, 5041 (1994)

200. S. Uhlenbrock, Investigations into the electronic structure of simple transition-metal oxides – with particular respect to nickel oxide (in German) "Untersuchungen zur elektronischen Struktur einfacher Übergangsmetall-Oxide – unter besonderer Berücksichtigung des Nickel-Oxids", Dissertation, University of Osnabrück, 1994

201. J. van Elp, R.H. Potze, H. Eskes, R. Berger, and G.A. Sawatzky, Electronic structure of MnO, Phys. Rev. B **44**, 1530 (1991)

202. J. van Elp, J.L. Wieland, H. Eskes, P. Kuiper, G.A. Sawatzky, F.M.F. de Groot, and T.S. Turner, Electronic structure of CoO, Li-doped CoO, and LiCoO$_2$, Phys. Rev. B **44**, 6090 (1991)

203. J. van Elp, H. Eskes, P. Kuiper, and G.A. Sawatzky, Electronic structure of Li-doped NiO, Phys. Rev. B **45**, 1612 (1992)

204. S. van Houten, Semiconduction in Li$_x$Ni$_{(1-x)}$O, J. Phys. Chem. Solids **17**, 7 (1960)

205. J. van Laar and J.J. Scheer, Influence of volume dope on Fermi level position at gallium arsenide surfaces, Surf. Sci. **8**. 342 (1967)

206. J.H. van Vleck, The puzzle of rare-earth spectra in solids, J. Phys. Chem. **41**, 67 (1937)

207. B.W. Veal and A.P. Paulikas, Final-state screening and chemical shifts in photoelectron spectroscopy, Phys. Rev. B **31**, 5399 (1985)

208. C.A. Ventrice Jr., D. Ehrlich, E.L. Garfunkel, B. Dillmann, D. Heskett, and H.-J. Freund, Metallic-to nonmetallic transitions of Na coadsorbed with CO$_2$ and H$_2$O on the Cr$_2$O$_3$(111)/Cr(110) surface, Phys. Rev. B **46**, 12892 (1992)

209. D. Venus and J. Kirschner, Momentum dependence of the Stoner excitation spectrum of iron using spin-polarized electron-energy-loss spectroscopy, Phys. Rev. B **37**, 2199 (1988)

210. C.S. Wang, Electronic structure, lattice dynamics, and magnetic interactions, in *High Temperature Superconductivity*, ed. by J.W. Lynn (Springer, New York, 1990), p. 122

211. G.K. Wertheim and S. Hüfner, X-Ray photoemission band structure of some transition-metal oxides, Phys. Rev. Lett. **28**, 1028 (1972)

212. T. Wolfram. R.E. Dewames, W.F. Hall, and P.W. Palmberg, Surface magnetization near the critical temperature and the temperature dependence of magnetic-electron scattering from NiO, Surf. Sci. 28, 45 (1971)

213. C. Xu, M. Hassel, H. Kuhlenbeck, and H.-J. Freund, Adsorption and reaction on oxide surfaces: NO, NO$_2$ on Cr$_2$O$_3$(111)/Cr(110), Surf. Sci. **258**, 23 (1991)

214. J. Zaanen, C. Westra, and G.A. Sawatzky, Determination of the electronic structure of transition-metal compounds: $2p$ x-ray photoemission spectroscopy of the nickel dihalides, Phys. Rev. B **33**, 8060 (1986)
215. J. Zaanen and G.A. Sawatzky, The electronic structure and superexchange interactions in transition-metal compounds, Can. J. Phys. **65**, 1262 (1987)
216. F.C. Zhang and T.M. Rice, Effective Hamiltonian for the superconducting Cu oxides, Phys. Rev. B **37**, 3759 (1988)

Index

Springer Tracts in Modern Physics